Spring/86

May 15/86.

# Microelectronics in Aircraft Systems

**Boeing 757**—an example of an aircraft utilizing new electronic display technology. By kind permission of British Airways.

# Microelectronics in Aircraft Systems

**E H J Pallett** TEng (CEI), AMRAeS, FSLAET

**Pitman**

PITMAN PUBLISHING LIMITED
128 Long Acre, London WC2E 9AN

PITMAN PUBLISHING INC
1020 Plain Street, Marshfield, Massachusetts 02050

*Associated Companies*
Pitman Publishing Pty Ltd, Melbourne
Pitman Publishing New Zealand Ltd, Wellington
Copp Clark Pitman, Toronto

© E H J Pallet 1985

**Library of Congress Cataloging in Publication Data**

Pallett, E. H. J.
   Microelectronics in aircraft systems.
   Bibliography: p.
   Includes index.
   1. Airplanes—Electronic equipment.
   2. Microelectronics.
   I. Title.
TL693.P35  1985       629.135′5       84–4992
ISBN 0–273–08612–X (U.S.)

**British Library Cataloguing in Publication Data**

Pallett, E. H. J.
   Microelectronics in aircraft systems.
   1. Airplanes—Electronic equipment
   2. Microelectronics
   I. Title
   629.135′5    TL693

ISBN 0–273–08612–X

All rights reserved. No part of this publication may be reproduced. stored in a retrieval system, or transmitted, in any form or by any means, electronic, mechanical, photocopying, recording and/or otherwise, without the prior written permission of the publishers. This book may not be lent, resold, hired out or otherwise disposed of by way of trade in any form of binding or cover other than that in which it is published, without the prior consent of the publishers. This book is sold subject to the Standard Conditions of Sale of Net Books and may not be resold in the UK below the net price.

ISBN 0 273 08612 X

Text set and printed in Great Britain at The Pitman Press, Bath

# Contents

Preface  ix

**1 Developments  1**
Printed circuits  2
Integrated circuits  10
Fabrication of ICs  12
Cards and modules  20

**2 Number systems and coding  27**
Decimal number system  27
Binary number system  28
Octal number system  30
Hexadecimal number system  32
Coding  33
Parity in codes  40
Arithmetic operations  41
Transmission of digital data  44

**3 Logic gates and circuits  47**
Gates and their symbols  49
Logic circuit equations  58
Karnaugh maps  59
Dual functions of logic gates  60
Fabrication of gates  61
Logic concepts and 'non-electronic' systems  61

**4 Logic devices  65**
Flip-flops  65
Shift registers  67
Clocks  69
Counters  69
Encoders and decoders  70
Multiplexers  70
Demultiplexers  72

Binary adders 73
Operational amplifiers 73

## 5 Displays 81
Matrix displays 82
Encoding and decoding 93
Counting 97
Frequency dividing 103
Synchronous counters 103

## 6 Cathode ray tube displays 105
Principle of the CRT 106
Colour CRT displays 109
Scan conversion 110
Screen format 110
Colour generation 112
Alphanumeric displays 115
Electronic instrument display systems 117

## 7 Logic diagrams and interpretation 135
Equations and logic diagrams 137
Boolean equations using signal functions 139
Examples of practical logic diagrams 141

## 8 Computers 149
Analog computers 149
Digital computers 150
Computer classification 152
Computer languages 152
Microprocessors 153
Data transfer 153
Memories 156
Data conversion devices 163
Computer applications 164

## 9 Aircraft systems 189
Altitude reporting system 189
Flight data recording 193
Flight data acquisition systems 199
VOR system 200
Multiplex systems 206
Servo-operated instruments 207

## 10 Handling of microelectronic circuit devices 213
Static electricity  213
Identification  216
Protection and packaging  216
Static-free work stations  217
Handling outside special handling areas  219

## Appendices 221
Appendix 1  Abbreviations associated with microelectronic components, circuits and systems  221
Appendix 2  Powers of 2  226
Appendix 3  Classification of integrated circuits  227
Appendix 4  Coding of integrated circuit packs  228
Appendix 5  Symbols associated with characteristics of solid-state devices  230
Appendix 6  Acronyms and abbreviations associated with avionic systems, equipment and controlling signal functions  232
Appendix 7  Circuit diagram symbols  240
Appendix 8  Computer languages  243
Appendix 9  Abbreviations and acronyms used in flight management system displays  244
Appendix 10  Aircraft utilizing electronic display technology  248

## Bibliography 253
Avionics and equipment  253
Computers  254
Digital techniques, data transfer and logic  254
Displays  254
Electronics  254
Equipment handling  255
Glossaries and terminology  255
Integrated circuits  255
Microprocessors  255
Printed circuits  256
Symbols and diagrams  256

## Exercises 257
Chapter 1  257
Chapter 2  257
Chapter 3  259
Chapter 4  261
Chapter 5  261
Chapter 6  263
Chapter 7  264
Chapter 8  265
Chapter 9  266

**Solutions to exercises** 267

Chapter 2   267
Chapter 3   269
Chapter 4   271
Chapter 5   271
Chapter 6   271
Chapter 7   272
Chapter 8   272

**Index** 273

# Preface

Throughout the development of electronics there have been a number of revolutionary changes but undoubtedly, the most notable of these was the one associated with solid-state electronics theory, which culminated in the universal adoption of semiconductor devices in systems dependent on the processing and transference of control data. Apart from the unique and wide-ranging signal processing capability such devices had to offer, the vast reduction in their physical dimensions was also highly significant in that a greater number of circuit elements could be assembled on associated printed circuit boards.

The subsequent developments of design and manufacturing processes made it possible to combine even larger numbers of passive and active circuit elements with interconnecting 'wiring' into one minute piece or 'chip' of solid-state material. Thus, 'microelectronic' integrated circuit packs capable of performing a vast number of individual dedicated functions, ranging from simple switching to full processing and computation of digital data, became a reality. This was particularly advantageous to the designers of systems requiring electronic means of control, because in knowing the operational parameters involved, and the functions constituent circuit elements were required to perform, the design of complete circuits could be based on the selection of appropriate 'off-the-shelf' items, and on the adoption of a 'building-block' technique. This, in turn, was to result in vast improvements in the modular concept of interconnecting circuits performing dedicated functions.

As far as the control of aircraft systems is concerned, the application of microelectronic circuit devices has been somewhat more gradual in its approach in comparison with other areas of science and engineering, particularly in respect of control by digital computer-based systems. Thus, for the maintenance engineer progressing his responsibilities into the technology of the current generation of aircraft, a basic understanding of 'microelectronics' and the operating principles of the many associated devices, has largely been derived from books concerned with applications other than aircraft.

Hundreds of such books have of course, been published since the birth of what has been aptly termed the 'microelectronic revolution', and despite the valuable information they contain, it has often been a little difficult for engineers to choose appropriate material which would supplement their study requirements connected say, with aircraft-type courses or licence examinations. This has led to the frequently expressed view that a more useful purpose could perhaps be served by a book which, in combining many of the fundamental principles with some representative examples of aircraft systems applications, would bring together much of the 'need-to-know' information within one set of covers. In response therefore, to what seemed an interesting challenge, I have formulated this book along these lines, and it is hoped that it will be of some value to engineers, whether specializing in the maintenance of avionic systems, airframes and systems, or engines, and to those whose chosen role is the important one of training such engineers.

Since a great deal of the preparatory work on the book coincided with the introduction into public transport service of what have been termed 'new technology aircraft' namely, Boeing 757, 767, and Airbus A310, as much detail of systems as possible within book size limitations has been included, particularly in respect of the CRT display systems employed in these aircraft.

The coverage of the principles of logic devices and circuits is presented at a depth considered sufficient to meet appropriate examination requirements, but for readers desirous of widening the coverage, a bibliography has been included. A large number of self-test questions are also provided together with appendices containing selected reference data.

In conclusion, I would like to express my grateful thanks to the following organizations for their assistance in supplying material for reference purposes, and for granting permission to reproduce photographs provided by them: Airbus Industrie, British Airways, British Caledonian Airways, British Airways Helicopters, British Aerospace PLC, Bendix Corporation, Boeing Airplane Co, CSE Aircraft Services, Finnair, Gulfstream Aerospace Corporation, Intel Corporation, Marconi Avionics Ltd, Newmarket Transistors Ltd, Normalair-Garrett Ltd, Smith's Industries PLC, and Texas Instruments.

Copthorne
Sussex

E.P.
1984

# 1 Developments

Early forms of systems based on electronic technology were designed principally to meet the requirements for radio communication, navigation and automatic flight control. These systems depended for their operation on electron tubes which, in various configurations, were used in conjunction with numerous discrete components such as resistors, capacitors and transformers for the purpose of signal rectification, amplification, discrimination and transmission of appropriate control signals. All of these components were compactly mounted on a metal chassis contained within a suitably designed casing. With the continuing development of aircraft, it also became opportune to widen the applications of systems using electron tubes in order to meet the demands of a corresponding increase in operational requirements. This, however, necessitated the addition of more and more circuitry, components and associated wiring; and without some system design changes, individual units such as radio receiver/transmitter units, autopilot amplifiers, etc., would have become unmanageably large and heavy.

Initially, and then only to a limited extent, this problem was alleviated by adopting miniaturized electron tubes and interconnecting wiring. An example of an autopilot amplifier which was in service in the late 1940s and early 1950s is shown in Fig. 1.1. It used twenty-one miniature electron tubes contained in a casing $560 \times 420$ mm ($22 \times 16.5$ in) and weighing nearly 43 pounds.

At about this time, two further developments were taking place which were to change significantly the principles and practices of electronic systems design, namely the development of printed circuits and of transistors. Printed circuits made possible the elimination of a considerable amount of conventional wiring, thereby reducing the bulk of system units, while transistors were designed to perform the same functions as electron tubes and thereby completely eliminate their use. Furthermore, the solid-state phenomena from which transistors stemmed made possible drastic reductions in component sizes and, perhaps most significant

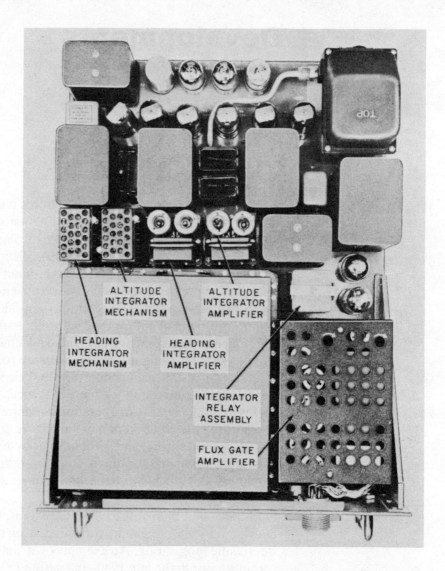

**Fig 1.1** Electron tube amplifier

of all, paved the way for the development of microelectronics and all associated integrated circuit elements.

**Printed Circuits**

Assembly of the various circuits which form the functional units of an avionic system necessitates the interconnection of individual components by means of electrical conductors. The commonest form of conductor is, of course, the single or stranded wire which interconnects components through soldered joints or screw-type terminal 'tags'. With the development of systems and their associated circuits, components became more numerous, with a corresponding increase in the number of required conductors and terminations. This problem was solved initially by using wire conductors of miniature dimensions and 'lacing' them into looms, as illustrated by the example of a 1949 vintage h.f. communication transmitter/receiver in Fig. 1.2. Although this method provided for a neat wiring layout and, to a certain extent, a form of modular build-up of sub-assemblies, the looms added considerably to the bulk of system units. Furthermore, considerable time had to be spent on making individual wire connections, particularly those involving soldering operations. It was desirable, therefore, for an alternative conducting method to be adopted, and so further development culminated in the technique whereby circuits could actually be printed and so eliminate a considerable amount of 'hard wiring'.

The printed circuit technique is one in which a systematic pattern of conducting paths is printed on a metallic foil bonded to an insulated baseboard. Since terminal points for the soldering of the circuit components are also provided, then as a single assembly a printed circuit board, as it is called, satisfies the structural and electrical requirements of the unit of which it forms a part.

**Board Construction**

The design of boards, and the production of complete assemblies, are of a specialized nature and both may vary according to individual circuit requirements and specifications. Boards are usually designed and manufactured in three basic configurations: single-layer, multilayer, and multilayer sandwich. Single-layer boards contain all printed conducting paths on one side with the discrete, or separate, components such as resistors, capacitors and transistors mounted on the opposite side. Multilayer boards have printed conducting paths on both sides, the transfer of circuits from one side to the other being accomplished by plated-through hole connections; discrete components may also be mounted on both sides of the board. Multilayer sandwich boards are actually many thin boards laminated together with the components on one, or sometimes both, of the external layers. There may be as many as 20 conductive layers in a multilayer sandwich board.

In construction, a board is essentially a rigid sheet of insulating base material to which a sheet of high-purity copper foil is bonded. The base material serves as a mounting for the discrete components which comprise the circuit, and is commonly made up either of layers of phenolic resin impregnated paper, or of epoxy resin impregnated glass-fibre cloth. The thickness of the base material depends on the strength and stiffness requirements of the finished

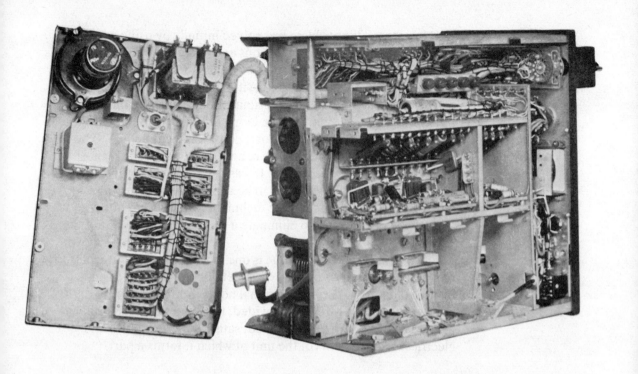

**Fig 1.2** Vintage (1949) h.f. communication transmitter/receiver

board, which in turn are dictated by the weight of the components to be carried, and by the size of the printed circuit area. Bonding of the copper foil is achieved by a hot-pressing operation which melts the resin in the base material so that it fully 'wets' the material and the copper foil. As polymerization of the resin mix proceeds, the base material reaches a fully cured state with the copper foil firmly bonded to it.

## Master Diagrams

The quality of a printed circuit board is, in the first instance, dependent on the production of enlarged scale master diagrams (Fig. 1.3) which must show precisely the circuit conductor pattern required, where all discrete components are to be located, circuit module designations, and other essential references. The production of diagrams requires the application of specialized

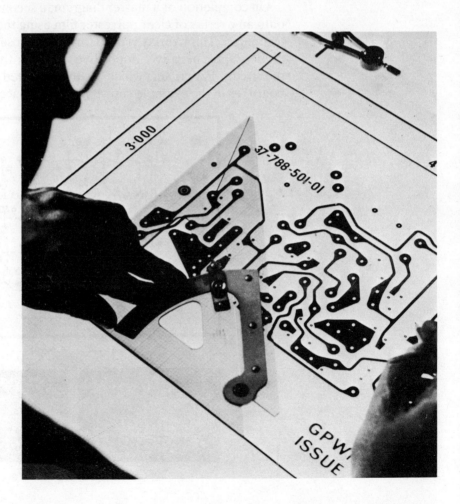

**Fig 1.3** Preparation of a master drawing for a printed circuit (courtesy Smiths Industries Ltd)

drafting techniques because, unlike conventional circuit and wiring diagrams which are drawn only as a guide to the build-up of an assembly of wiring and component connections, a printed circuit board must be an exact 'metallic' copy of the diagrams produced for it.

Diagrams are normally produced by adopting computer-aided design techniques, but the basic principle may be understood by considering the 'manual drawing' technique which is still in use. In this technique, black self-adhesive material produced in 'tape' form to represent conductors, and in other shapes to represent terminal points, edge connector contacts, drilling points, component outlines, etc, is applied to a sheet of polyester film in accordance with the required pattern layout. The material is produced in a wide range of sizes to suit both the scale selected for the artwork overall, and the reduction ratio required for the subsequent processes of photographing and printing onto the metallic foil.

On completion of a master diagram, a second master is copied onto an overlay of clear polyester film using the same 'drawing' technique. This overlay is used for the purpose of producing a photographic negative, or positive, plate for the actual printing of the circuit. For circuits which are to be printed on both sides of a board, as in the example shown in Fig. 1.4, accurate registering of

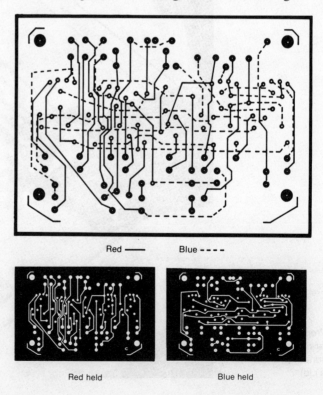

**Fig 1.4** Double-sided circuit master

the circuit patterns and terminal connections during the photographic and printing stages is essential. This is ensured by superimposing both patterns on a single film overlay master diagram and differentiating between them by applying coloured self-adhesive transparent tapes, e.g., blue tapes corresponding to one side and red tape to the opposite side At the photographic stage, coloured filters are used to separate the circuit artwork into its two individual patterns and with the terminal areas, datum points, etc., in perfect register with each other.

**Printing of Circuits**

The printing process is a photolithographic one involving the use of a photographic positive (negative in some cases) of the master diagram of the circuit, and a photosensitive material which has the property of becoming soluble when exposed to strong light. The solution, which is known as a 'resist', is coated onto the cleaned surface of the copper foil (Fig. 1.5(a)) and the photographic positive

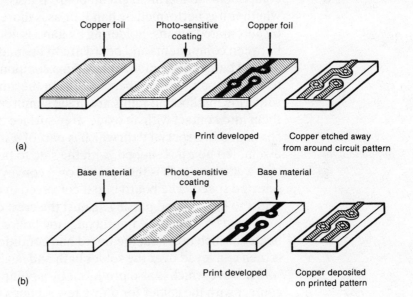

**Fig 1.5** Printing processes: (a) Etching. (b) Additive

of the diagram is placed over the board and time-exposed in a printing machine. After exposure, the resist is washed away to leave unprotected areas of copper around the circuit pattern, and the board is dried by a clean air blast. The board is then placed in a bath containing an etching solution, such as ferric chloride or ammonium persulphate, which etches away all the unprotected copper. When the etching process has been satisfactorily completed, the board is thoroughly washed in water to remove all traces of etching solution, and is then dried.

As an alternative to the etching process, printing may be done by an 'additive' process, i.e., a process by which copper is deposited only in the areas where conducting paths are required (Fig. 1.5(b)). The base material of the board is pre-coated with a specified adhesive, circuit holes are pre-fabricated, and the board is sensitized by a coat of resist solution. A photographic negative of the conducting pattern is then printed on the board so that the exposed areas define the conductor network. These exposed areas are chemically activated, and the board is then immersed in a copper plating solution. After a period of time consistent with the deposition of the required thickness of copper, the board is removed from the solution.

### Soldering of Boards and Components

The technique of soldering the requisite discrete and active components to a printed circuit board is also a specialized one, whether it is performed by hand, or as is more generally the case, by automated means. Soldering by hand is adopted where joints between components and board are to be made separately, e.g. in limited batch production, and when a component is replaced after a test or a repair has been carried out. In the automatic or mass soldering method, all joints are made simultaneously by bringing them into contact with an oxide-free surface of molten solder contained in a special bath which is part of a soldering machine. The assembled board is clipped, with the side to be treated downwards, into a carrier which is then placed on a conveyor for traversing at a selected speed. The board is first conveyed over a fluxer unit, the joints to be soldered passing against the crest of a wave of flux produced by a pump. After fluxing, the board passes over radiant heaters which pre-heat the board and condition the flux. The board is then conveyed over the solder bath and against the crest of a solder wave which is also produced by a pump. Each joint area is in contact with the solder for only a few seconds to prevent distortion and damage to the board and its assembled components.

Other automated processes in use are reflow or heat-cushion soldering, and laser jointing.

### Flexible Printed Circuits

A flexible printed circuit is basically a system of copper foil conductors thermally bonded to a base of thin flexible insulator and

**Fig 1.6** Example of PCBs applied to an engine control unit. A flexible circuit interconnection is shown on the extreme right of the picture (courtesy Smiths Industries Ltd)

used to interconnect (see example shown in Fig. 1.6) separate assemblies and devices in electronic equipment, particularly those which may be moved relative to each other, and those which may be mounted in different planes. They can also permit easier assembly and higher density packaging of units. The base material is either polyester film or polyimide film (normally used in applications to aircraft systems) and the conductors are bonded to it by modified epoxy or an acrylic type adhesive. The conductors are encapsulated by a thermally bonded insulating overlay of the same material as the base.

Although flexible circuits are often used to replace a wiring harness or loom, a direct comparison between the two could be misleading. A flexible circuit is a true repeatable component in its own right, albeit custom-made, and as such may be obtained as a replacement item against a specification in much the same way as with other discrete components.

**Integrated Circuits**  Like transistors, integrated circuits (ICs) are also based on solid-state electronic technology. However, whereas independent transistors require point-to-point wiring and interconnection with discrete passive components such as resistors and capacitors, in integrated circuits active and passive components, and conducting paths, can all be fabricated within a single structure of the solid material; hence their name. By this method of fabrication it is therefore possible to provide 'packaged' microelectronic devices which perform all the functions associated with the processing of data in both linear and digital form (see also Appendix 3).

ICs may be categorized according to the manufacturing processes employed for their fabrication, and to their circuit function. General descriptions related to the manufacturing processes include thin- and thick-film, fully integrated or monolithic, multichip, and hybrid. Other descriptive adjectives are often appended to specify the technology and materials of particular circuits in more detail, e.g., planar, diffused, junction isolated, silicon monolithic ICs.

**Film Circuits**

In film circuits, all components are fabricated by depositing passive or active layers onto a single common supporting insulating layer or *substrate*, conducting interconnections being made by a superimposed metallic film. There are two versions of film circuits: thin-film and thick-film. In the *thin-film* version the substrate is often a ceramic, or a glass slice, e.g., alumina, and the layers, which are only a few micrometres thick, are deposited by vacuum evaporation, cathode sputtering, or other thin-film techniques. In the *thick-film* version the various layers (up to about 20 micrometres thick) which constitute each component (usually passive) and its interconnections, can be deposited onto a substrate by means of a screen printing process. Thick-film circuits may also be used to form hybrid ICs.

## Monolithic ICs

These circuits are so called because all the active and passive components are formed as an integral part of a single chip of semiconducting substrate, normally silicon. The active layers in the silicon which constitute each component are produced by modifying the material by doping successive layers with impurities, the dopant being introduced by diffusion or epitaxial techniques. This produces, for example, n-p-n bipolar transistor structures and derivative components in the silicon. Interconnections between the various parts of the circuit are made by depositing a pattern of conducting material, such as aluminium, on an oxide insulating surface. This is the basis of the widely used technology known as MOS (metal-oxide-silicon).

Because each circuit and its components can be made so small, many identical circuits can be fabricated simultaneously on one slice of silicon, or alternatively many different types of circuit can be made on a slice and interconnected with a metallizing layer to produce a complete microelectronic sub-system. Some idea of the small dimensions involved can be gained from Fig. 1.7. Assuming a

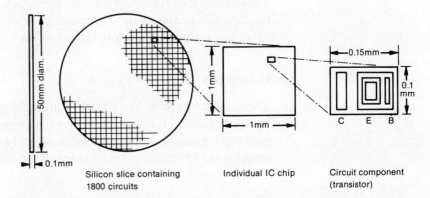

**Fig 1.7** Typical geometry of silicon slice and IC components

silicon slice is of 50 mm diameter, then it can accommodate approximately 1800 ICs each typically occupying a 1 mm square chip of material. A bipolar transistor, which may be an element of a circuit, occupies an area of (say) 0·15 mm × 0·1 mm, and other elements can be smaller; and so from the dimensions assumed, it is possible to have 50 elements on each IC. An increase in slice diameter will obviously increase the number of elements possible. Furthermore, by the use of MOS technology, or a bipolar process known as collector-diffused isolation (CDI), the areas of devices can be reduced, so that a corresponding increase in packing density of components and circuits can be obtained. In connection with packing density it is usual to categorize ICs in terms of hardware complexity or scale of integration, and so at this point it is pertinent to define the categories.

*Scale of Integration*

In categorizing the scales of integration, it is usual to adopt the gate or equivalent circuit as the unit. There are four codes as follows:

- Small-scale integration (SSI)—containing not more than 11 gates per IC
- Medium-scale integration (MSI)—containing between 12 and 99 gates per IC
- Large-scale integration (LSI)—containing between 100 and 1000 gates per IC
- Very-large-scale integration (VSLI)—containing over 1000 gates per IC

**Multichip ICs**

These consist of monolithic circuits, or parts of circuits, which are bonded to a common substrate and interconnected by bonded gold wires to produce a complete circuit or sub-system.

**Hybrid ICs**

These circuits use thin- or thick-film techniques to form passive components and interconnections on an insulating substrate, but the active components are produced separately and designed for direct attachment to the interconnections of the film circuit.

**Fabrication of ICs**

The structure of an IC is complex both in the topography of its surface and in its internal composition; each element having an intricate three-dimensional architecture that must be reproduced exactly in every circuit and in each layer. The fabrication process is therefore a very sophisticated one requiring precise interpretation of its designed plan.

**Fabrication of monolithic ICs**

In some respects, fabrication is not unlike that of printed circuit boards, in that it is dependent in the first instance on basic circuit layout diagrams from which highly accurate master artwork

**Fig 1.8** Preparation of master for a MOS chip (courtesy Smiths Industries Ltd)

(Fig. 1.8) can be generated, and checked, by computerized control methods. The resulting layout is then used to prepare a set of photomasks (Fig. 1.9) each containing the pattern for a single layer. The masks are checked to verify that every circuit element is correctly placed and of the right size, and this is usually done by first generating an enlargement of each layer (typically 10×) on a glass photographic plate called a reticle, and then viewing a further enlargement of this. Typically this enlargement would be 500 times that of the actual circuit. Master masks are then prepared by projecting a reduced image of the reticle onto another photographic plate by means of a lens system capable of extremely high resolution, and from these a number of working plates are copied. An alternative method of producing masks is the electron-beam

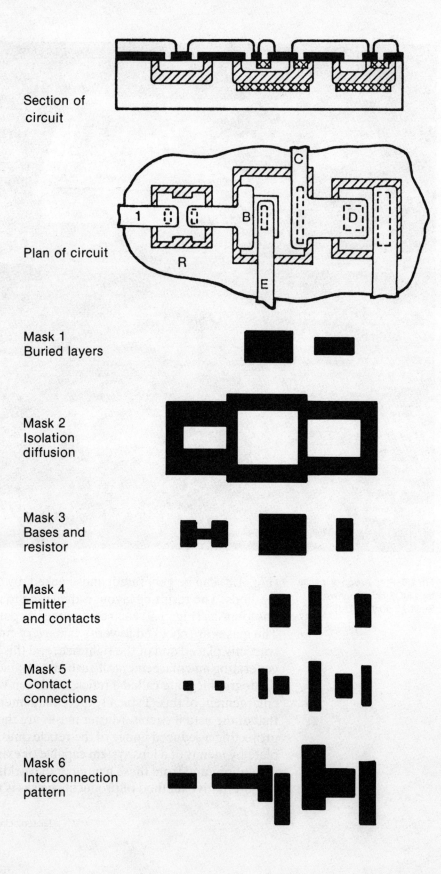

Fig 1.9 Masking

lithographic method enabling patterns to be 'written' directly on the working mask from information stored in the design computer memory.

Figures 1.10(a)–(k) illustrate the various stages of fabrication of a monolithic IC, and although this is based on a hypothetical and simple circuit, it should convey an understanding of the processes typically involved.

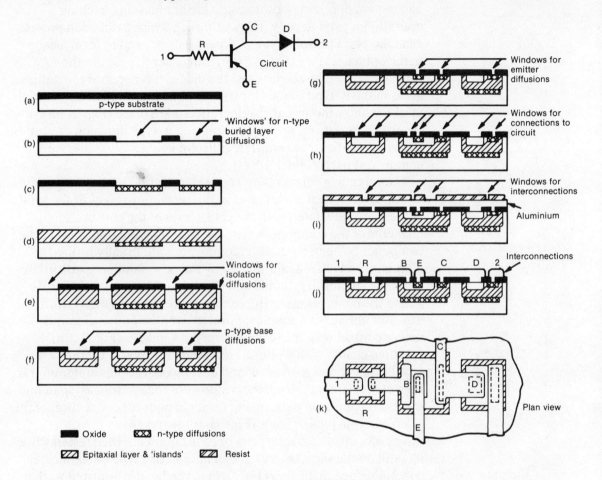

**Fig 1.10** Stages in fabrication of a monolithic IC

The primary part of the IC is the *substrate*, and this is a slice of almost pure silicon which has been sawn from a 'grown crystal' bar of the material. The slice is doped to form a p-type semiconductor and polished to a mirror-like finish. Typically, the dimensions would be as already indicated in Fig. 1.7. An oxide layer is formed on one surface of the slice, and through this layer 'windows' are etched by a photolithographic process as in Figs. 1.10(a) and (b) at the positions pre-determined by the photographic masks. The purpose of the windows is to permit heavily doped n layers to be

Microelectronics in Aircraft Systems 15

diffused within the substrate (Fig. 1.10(c)) in the regions where the elements of the active component, i.e., the emitter and collector regions of the transistor, are located. The oxide layer is subsequently removed, and the diffusions are then buried by a thin layer (typically 10 $\mu$m thick) of n-type silicon which is grown on the substrate. This process is known as *epitaxy*\* and the layer is referred to as the *epitaxial layer* (Fig. 1.10(d)). This layer is then oxidized and, with a different photomask, a further masking/etching operation is used to cut windows through which a diffusion process can take place to isolate the p-type substrate, and to form 'islands' in the epitaxial layer (Fig. 1.10(e)). In order to produce the transistor base, the resistor and the diode, it is necessary to define areas in the n-type islands for p-type diffusions, and so yet another masking and window cutting operation takes place (Fig. 1.10(f)). A further oxidation, window cutting and n-type diffusion operation is carried out to create emitter, collector contact areas and similar component parts (Fig. 1.10(g)).

The foregoing stages complete the fabrication of the components of our hypothetical circuit, and so the remaining stages of the total process are associated with the connection of the constitutional elements of the components and the overall interconnection of the components themselves. Interconnections are usually formed in aluminium which is again shaped by the now-familiar oxidizing, masking, photolithographic and window-cutting operations to define the connections to the component elements (Fig. 1.10(h)). The aluminium is vacuum evaporated over the entire slice and makes contact with the elements via the windows (Fig. 1.10(i)). It is then coated with a photoresist which is exposed via the appropriately designed mask and developed. Surplus aluminium is then etched away through the windows of the photoresist to define the interconnection pattern and terminal pads for connection of the IC to external pins or leads (Figs. 1.10(j) and (k)).

The electrical characteristics of each circuit are then tested while it is still on the slice, by means of a microscope type of testing head containing needle probes (Fig. 1.11). The head is adjusted so that the probes contact the aluminium terminal pads of each circuit in turn, thereby connecting them to the test measuring instruments and power supplies. For complex circuits, the testing operation is computer controlled. On satisfactory completion of the tests, individual circuits are separated from each other by diamond tool scribing and cleaving (Fig. 1.12) to produce small chips of complete ICs. Finally, each chip is fixed to a suitable base or header, and fine

---

\*From the Greek word meaning 'arranging upon'.

**Fig 1.11** Close-up of probe testing on slice (courtesy Smiths Industries Ltd)

gold wires are bonded to the circuit terminal pads and header lead-throughs, to form the ultimate connection between the IC and the 'outside world'. The circuit and header are encapsulated in a thermosetting epoxy plastic. This packaging process is then followed by final quality control checks.

The choice of a suitable package is wide and is dependent on the application and quality of the finished product. Three packages are in common use: (i) the *flat pack*, so called because the leads are soldered flat onto the conductors of a printed circuit board, which is usually hermetically sealed; (ii) the '*can*' pack sealed by a metal cover spot-welded to the header; and (iii) the *dual-in-line* (DIL) pack which is encapsulated in a thermosetting epoxy.

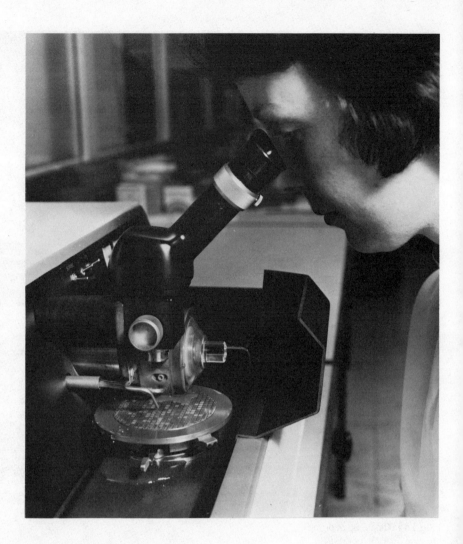

**Fig 1.12** Sawing a silicon slice to separate individual circuits (courtesy Smiths Industries Ltd)

### Fabrication of Hybrid ICs

The fabrication of hybrid ICs is different from that for monolithic ICs in that, firstly, the substrate is a flat ceramic material (alumina) onto which conductive patterns and resistors are printed by means of pastes or 'inks'; and secondly, that other discrete active or passive components designed in accordance with one or other of the IC techniques are wire-bonded or reflow-soldered to the conductors. There is, therefore, a fundamental similarity in concept between a hybrid IC and a printed circuit board and its components, and this may be seen from Fig. 1.13 which also illustrates yet again the vast reduction in overall dimensions made possible by microelectronic technology.

Not unlike the monolithic IC fabrication, the individual IC

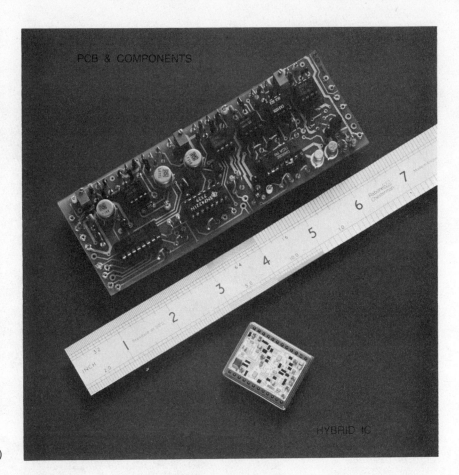

**Fig 1.13** Comparative size reduction between PCBs and a hybrid IC (courtesy Newmarket Microsystems Ltd)

patterns are set out simultaneously on a single piece of substrate, which is in this case, alumina and typically 50 mm square. The printing pastes are finely ground mixtures of glass and noble metals, e.g., platinum or palladium-silver for conductors, and glass and noble metal oxides for resistors. Printing is by means of an automatic screen printing machine, and following this process the pastes are dried and fired in a special furnace at 850°C to fuse the glass onto the substrate. Different values of resistors are obtained by altering the length-to-width ratio of the resistive patterns, and after printing, individual resistors are trimmed to the required value by a computer-controlled laser beam trimming head (see Fig. 1.14). Multilayer circuits can be built-up on the substrate by printing dielectric layers over the lowest conductor pattern and adding further printed conductor patterns above it. Interconnection between appropriate points of the conductors is made via holes in the dielectric. Separation of the individual circuits from the

**Fig 1.14** Laser beam trimming head (courtesy Newmarket Microsystems Ltd)

substrate is achieved by scribing the substrate with a laser beam and then snapping the pieces off.

Discrete components are added to the substrate to complete the circuit. The components can be either unpackaged semiconductor planar chips, or pre-packaged IC devices. Unpackaged planar chips are attached to the substrate by means of either a silver-based or gold-based epoxy, and then by gold wire ultrasonically bonded to the top surface of the chip and the appropriate conductor tracks (Figs. 1.15 and 1.16). Further examples of hybrid IC packaging are shown in Fig. 1.17, and it will be noted that single-in-line (SIL) pin configurations are used as well as DIL.

**Cards and Modules**

In the majority of avionic systems the circuitry is very complex, and so there is obviously a proportionate increase in the number of

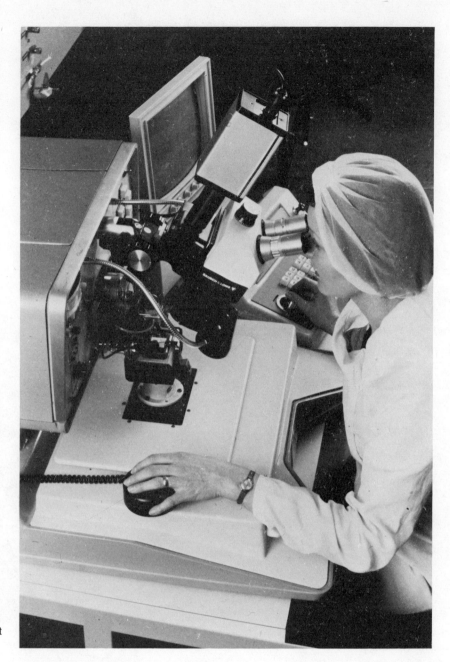

**Fig 1.15** Automatic gold wire bonding (courtesy Newmarket Microsystems Ltd)

components to be accommodated on printed circuit boards. In such cases it is usual for the overall circuit to be designed in a 'modularized' form so that a number of boards can be assembled each having a specific function to perform, and of such dimensions that the volume available within the confines of the various standard

**Fig 1.16** Example of hybrid IC (courtesy Newmarket Transistors Ltd)

**Fig 1.17** Various hybrid IC packages and encapsulations (courtesy Newmarket Transistors Ltd)

ARINC case configurations can be utilized to the fullest effect.

The boards, which are generally referred to as 'cards' or 'card modules', may be of the double-sided and/or multilayer types, and when inserted in their allotted positions within a case their circuits are interconnected either by automatic 'mating' with edge-type

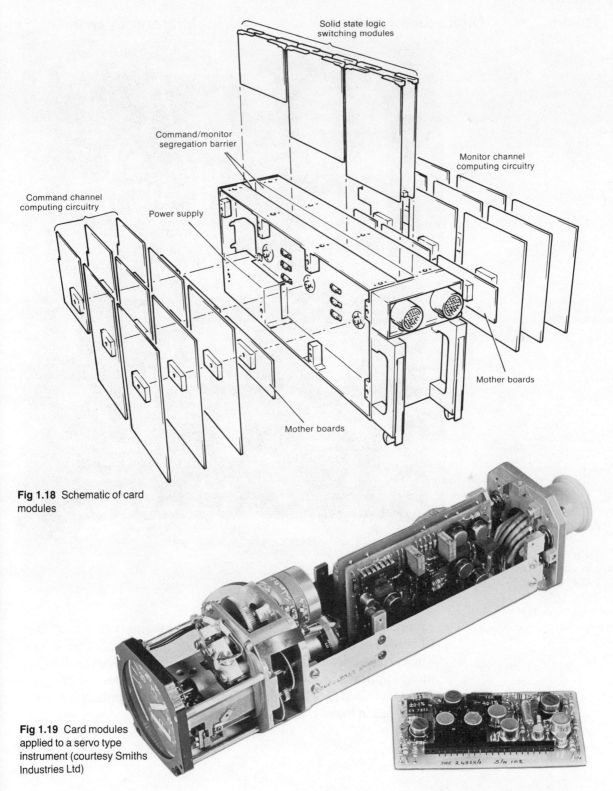

**Fig 1.18** Schematic of card modules

**Fig 1.19** Card modules applied to a servo type instrument (courtesy Smiths Industries Ltd)

Microelectronics in Aircraft Systems

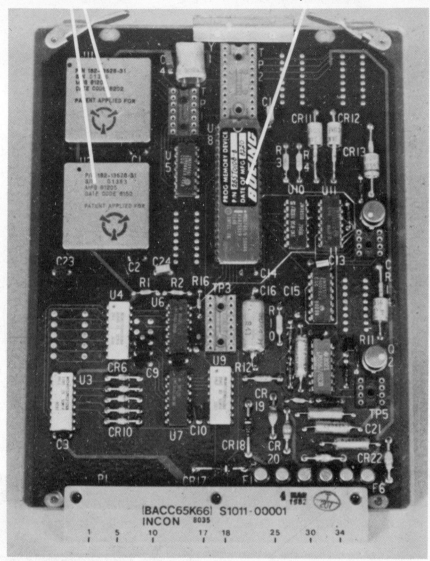

**Fig 1.20** Multilayered digital circuit card (courtesy Boeing)

connectors secured to the main frame of the casing, or by coupling ribbon-type cables to connectors mounted on the cards themselves. The cards are secured to the main frame by fixing screws and, as in many applications, by quick-release extractor handles which facilitate removal of a module. Some examples of applications are illustrated in Figs. 1.18 to 1.21.

**Fig 1.21** Card modules of an autopilot pitch-axis computer (courtesy Marconi–Elliott Avionic Systems Ltd)

**Boeing 767**—an example of an aircraft utilizing new electronic display technology. By courtesy of Dick Kerry.

# 2 Number systems and coding

Most digital equipment deals with numbers, the actual forms of the input and output numbers being dependent on the applications of the equipment. They may be in decimal form, binary form, octal or hexadecimal forms, and in some applications, the input and/or output may be in analog form, i.e., continuously variable, despite the digital processing.

**Decimal Number System**

This system, which is the most familiar to us, is one in which the ten digits 0 to 9 are combined in a certain way so that they indicate a specific quantity. The basic distinguishing feature of this system, and in fact in all number systems adopted in digital equipment, is its *base* or *radix* since it indicates the number of characters or digits used to represent quantities in the system. Thus, the decimal system has a base or radix of 10 because we use the ten digits 0 to 9. The number 10 is not a basic digit, since it is the result of the digits 1 and 0 and so is specifically '1 ten' and '0 units'.

Each digit position in a decimal number carries a particular weight in determining the magnitude of that number, e.g., weights of units, tens, hundreds, thousands, etc. The weight is some *power* of the radix 10 and is indicated by a number called an *index* or *exponent* written as a subscript so that it tells us how many times 10 is to be multiplied by itself when using it as a factor. Thus $10^3$ is read as 'ten to the power of 3' and is equal to $10 \times 10 \times 10$ or 1000. Consider, for example, the number 3584; we have 4 units, 8 tens, 5 hundreds and 3 thousands. The number can therefore be written as:

$$(3 \times 10^3) + (5 \times 10^2) + (8 \times 10^1) + (4 \times 10^0)$$
$$= 3000 + 500 + 80 + 4$$
$$= 3584$$

In any decimal number the digit having the smallest weight is known as the *least significant digit* (LSD) and the one having the greatest weight or value is known as the *most significant digit*

(MSD). In the above example, the corresponding digits are 4 and 3 respectively.

**Number Identification**

When we are working with more than one number system, it is often necessary to identify the radix of each number; this is particularly so with numbers which include zeros and ones. For example, the number 101 could represent a quantity of 'one hundred and one' in decimal, but in binary form it would represent a quantity of 'five'. A small subscript number (not to be confused with the index or exponent referred to earlier) is therefore written after the principal number to indicate its system radix. Thus, the decimal system radix being 10 and the binary system radix being 2, the notation for the example chosen would be:

$$101_{(10)} = 1100101_{(2)} \quad \text{and} \quad 101_{(2)} = 5_{(10)}$$

Examples of the notation related to the other number systems in use will be given later in this chapter.

## Binary Number System

The distinguishing feature of this system, on which all digital processing is dependent, is that it uses the radix 2. The two binary digits, or *bits* as they are always called, are 0 and 1, and when appropriately arranged can also represent any decimal number. Each bit position in a binary number carries a specific weight and the position weights are some power of the radix 2. Table 2.1 is a comparison between the decimal and binary systems.

To convert from decimal to binary, a continuous process of dividing the decimal number and quotients by 2 is carried out; the

**Table 2.1** Comparison between binary and decimal systems

| Binary | Decimal |
| --- | --- |
| $2^0 = 1$ | $10^0 = 1$ |
| $2^1 = 2$ | $10^1 = 10$ |
| $2^2 = 4$ | $10^2 = 100$ |
| $2^3 = 8$ | $10^3 = 1000$ |
| $2^4 = 16$ | $10^4 = 10\,000$ |
| . | . |
| . | . |
| . | . |

equivalent bits are then determined on the basis of whether or not there is any remainder from the division. The method of converting $30_{(10)}$ is given as an example in Table 2.2. Reading the last

**Table 2.2** To show that the binary equivalent of 30 is 11110

| $\dfrac{\text{Decimal number}}{\text{binary radix}}$ = | Quotient | Remainder or bit | |
|---|---|---|---|
| $\dfrac{30}{2}$ = | 15 | 0 | LSD |
| $\dfrac{15}{2}$ = | 7 | 1 | |
| $\dfrac{7}{2}$ = | 3 | 1 | |
| $\dfrac{3}{2}$ = | 1 | 1 | |
| $\dfrac{1}{2}$ = | 0 | 1 | MSD |

remainder as the MSD and the first remainder as the LSD, we note that the binary equivalent of $30_{(10)}$ is $11110_{(2)}$.

The binary equivalents of the decimal numbers 0 to 15 are given in Table 2.3, and as an exercise the reader may select numbers to

**Table 2.3** Binary equivalents of decimal 0–15

| Decimal | Binary | Decimal | Binary |
|---|---|---|---|
| 0 | 0000 | 8 | 1000 |
| 1 | 0001 | 9 | 1001 |
| 2 | 0010 | 10 | 1010 |
| 3 | 0011 | 11 | 1011 |
| 4 | 0100 | 12 | 1100 |
| 5 | 0101 | 13 | 1101 |
| 6 | 0110 | 14 | 1110 |
| 7 | 0111 | 15 | 1111 |

check understanding of the conversion method just described.

To convert binary to decimal it is only necessary to add the value for each position of a number having a bit equal to 1, as shown in Table 2.4.

**Table 2.4** Converting binary to decimal

| | $2^4$ | $2^3$ | $2^2$ | $2^1$ | $2^0$ | Base with exponent |
|---|---|---|---|---|---|---|
| | 16 | 8 | 4 | 2 | 1 | Decimal value of each position |
| | | | | | 1 | $= 1$ |
| | | | 1 | 0 | 1 | $= 1 + 4 = 5$ |
| | 1 | 1 | 0 | 1 | 1 | $= 1 + 2 + 8 + 16 = 27_{(10)}$ |

**Binary Number Sizes**

Binary numbers made up of the appropriate group of bits are also referred to as binary *words*, and the group of bits is termed a *byte*. Most digital circuits and equipment use a fixed word size, the size determining the maximum magnitude and resolution with which numbers can be represented. The number of bits in a word determines the number of discrete states that can exist and the maximum decimal number that can be represented. The formula used in this connection is $N = 2^n$, where $N$ is the total number of states and $n$ is the number of bits. Referring once again to Table 2.3, we can see that the binary numbers shown constitute a 4-bit word; thus, from the above formula a total of $N = 2^n = 2^4 = 16_{(10)}$ binary bit patterns or number combinations can in this case be created as shown in the table.

The largest decimal number value ($N$) that can be represented for a given number of bits ($n$) is, in fact, one less than the total number of states, and so it is expressed by $N = 2^n - 1$. Thus, for a 4-bit word, the maximum decimal value is $16 - 1 = 15$.

Appendix 2 contains a table of numbers that are powers of 2, and this will be found useful in determining the relationship between decimal numbers and binary word length.

## Octal Number System

The octal number system is often used in digital computer input and output units, and is convenient for representing binary numbers because it requires far fewer digits than does the binary system. The radix of the octal system is 8 and octal counting proceeds from 0 to 7 just as in the decimal system. The digits 8 or 9 do not exist in octal, and to progress from 7 requires a carry operation and use of a position value. Table 2.5 shows octal numbers and their decimal equivalents.

The conversion from decimal to octal follows the same rules as for decimal to binary, except that 8 is used instead of 2. The method of converting the number $3844_{(10)}$ to its octal equivalent $7404_{(8)}$ is shown as an example in Table 2.6.

**Table 2.5** Octal equivalents of decimal 0–10

| Octal | Decimal |
|---|---|
| 0 | 0 |
| 1 | 1 |
| 2 | 2 |
| 3 | 3 |
| 4 | 4 |
| 5 | 5 |
| 6 | 6 |
| 7 | 7 |
| 10 | 8 |
| 11 | 9 |
| 12 | 10 |

**Table 2.6** Demonstration that the octal equivalent of $3844_{(10)}$ is $7404_{(8)}$

| $\dfrac{Decimal}{Octal\ radix}$ = | Quotient | Remainder (octal digit) | |
|---|---|---|---|
| $\dfrac{3844}{8}$ = | 480 | 4 | LSD |
| $\dfrac{480}{8}$ = | 60 | 0 | |
| $\dfrac{60}{8}$ = | 7 | 4 | |
| $\dfrac{7}{8}$ = | 0 | 7 | MSD |

An example of converting an octal number to its decimal equivalent is given below:

$$227_{(8)} = (2 \times 8^2) + (2 \times 8^1) + (7 \times 8^0)$$
$$= (2 \times 64) + (2 \times 8) + (7 \times 1)$$
$$= 151_{(10)}$$

A requirement often exists for octal to binary conversion, and this is based on a substitution method which takes advantage of a natural relationship between octal and binary numbers. Since the radix 8 equals $2^3$, the relationship is that one octal digit may be expressed by three bits, as shown in Table 2.7. If, for example, we wish to convert $23_{(8)}$, then by direct substitution of each digit we note from

**Table 2.7** 3-bit binary equivalents of octal 0–7

| Octal | Binary |
|-------|--------|
| 0 | 000 |
| 1 | 001 |
| 2 | 010 |
| 3 | 011 |
| 4 | 100 |
| 5 | 101 |
| 6 | 110 |
| 7 | 111 |

the table that $23_{(8)}$ equals $010\ 011_{(2)}$. A check on the equivalence can be made by converting each to a decimal number; in this case, both would equal $19_{(10)}$. In the absence of a table, the conversion from octal to binary is carried out in the same manner as decimal to binary (see Table 2.2) except that division is by the radix 2.

Conversion from binary to octal is also done by substitution. The binary digits are arranged in groups of three, filling out the extreme left or right group with zeros if necessary, and then directly substituting the octal equivalent for each group. For example, in converting $11100_{(2)}$, since there are only five digits, the first two digits would be filled out with a 0; thus

$$11100_{(2)} = 011\ \ 100$$
$$= \ \ \ 3\ \ \ \ \ 4$$
$$= 34_{(8)}$$

## Hexadecimal Number System

This system is also used in digital computer input and output units, and is one in which the numbers are to the radix of $16_{(10)}$. It is unique in that the ten decimal digits 0 to 9 are used, together with letters A to F to represent 10 to 15. This is shown in Table 2.8.

**Table 2.8** Hexadecimal number system

| Decimal | Hexadecimal | Binary | Decimal | Hexadecimal | Binary |
|---------|-------------|--------|---------|-------------|--------|
| 0 | 0 | 0000 | 8 | 8 | 1000 |
| 1 | 1 | 0001 | 9 | 9 | 1001 |
| 2 | 2 | 0010 | 10 | A | 1010 |
| 3 | 3 | 0011 | 11 | B | 1011 |
| 4 | 4 | 0100 | 12 | C | 1100 |
| 5 | 5 | 0101 | 13 | D | 1101 |
| 6 | 6 | 0110 | 14 | E | 1110 |
| 7 | 7 | 0111 | 15 | F | 1111 |

Another feature of the hexadecimal system is that each digit is equivalent to 4 bits.

The conversion from decimal to hexadecimal is done by repeated division by $16_{(10)}$ as shown by the example in Table 2.9. Conversion from hexadecimal to decimal is similar to conversion from octal to decimal, the positional values being from right to left. For example, $23_{(16)}$ is equivalent to $35_{(10)}$ as follows:

$$
\begin{aligned}
23_{(16)} &= (2 \times 16^1) + (3 \times 16^0) \\
&= (32) + (3) \\
&= 35_{(10)}
\end{aligned}
$$

As another example:

$$
\begin{aligned}
1FF_{(16)} &= (1 \times 16^2) + (F \times 16^1) + (F \times 16^0) \\
&= (256) + (240) + (15) \\
&= 511_{(10)}
\end{aligned}
$$

Conversion from hexadecimal to binary, and vice versa, is the same as the substitution method adopted for octal/binary/octal conversions except that digits are grouped into four bits (see Table 2.9). For example:

$$
\begin{aligned}
1A6_{(16)} &= 0001\ 1010\ 0110 \\
&= 000110100110_{(2)}
\end{aligned}
$$

Similarly:

$$
\begin{aligned}
011011110101_{(2)} &= 0110\ 1111\ 0101 \\
&= 6F5_{(16)}
\end{aligned}
$$

**Table 2.9** Converting decimal to hexadecimal

| Decimal / Radix | | Quotient | Remainder | |
|---|---|---|---|---|
| $\dfrac{158}{2}$ | = | 9 | 8 | LSD |
| $\dfrac{9}{2}$ | = | 0 | 9 | MSD |
| | $158_{(10)} = 98_{(16)}$ | | | |

**Coding**

By coding we mean, of course, the method by which alphanumeric data can be specified by means of other symbols which can in turn be easily handled and transmitted by equipment dependent on digital processing techniques. We have already touched on the subject in dealing with number systems; for example, in converting from decimal to binary we have seen that binary equivalents consist of four or more bits, and so it can be stated that the binary equivalents are the 'pure binary code' for the decimal numbers. Similarly, the octal and hexadecimal number systems produce their appropriate codes. However, pure binary notation codes are somewhat restricted in their application, and so in the development of digital processing techniques and applications it also became necessary to develop other codes which would be more compatible with the specialized nature of the systems developed. Details of some of these codes are given in the following sections.

**Binary Coded Decimal (BCD)**

Many digital devices, instruments and equipment use decimal inputs and outputs, and so this code is widely used for the reason that it combines some of the characteristics of both the binary and decimal number systems.

The code uses a 4-bit binary number to specify the digit decimal numbers 0 to 9, and as indicated in Table 2.10 the weighting of a bit

**Table 2.10** 4-bit binary equivalents of decimal 1–9

| Decimal | BCD | | | |
|---|---|---|---|---|
| | $2^3$ | $2^2$ | $2^1$ | $2^0$ |
| | 8 | 4 | 2 | 1 |
| 1 | 0 | 0 | 0 | 1 |
| 2 | 0 | 0 | 1 | 0 |
| 3 | 0 | 0 | 1 | 1 |
| 4 | 0 | 1 | 0 | 0 |
| 5 | 0 | 1 | 0 | 1 |
| 6 | 0 | 1 | 1 | 0 |
| 7 | 0 | 1 | 1 | 1 |
| 8 | 1 | 0 | 0 | 0 |
| 9 | 1 | 0 | 0 | 1 |

position reading from left to right is 8–4–2–1 (see also Table 2.4), and for this reason the code is also called an '8–4–2–1 code'.

It will be noted that the 4-bit binary numbers are the same as those in pure binary notation (see Table 2.3), but here the similarity ends. In order to represent decimal numbers from 10 and upwards in BCD notation, the conversion is done simply by substituting the

4-bit code appropriate to each of the decimal digits 0 to 9. For example, $10_{(10)}$ in BCD is 0001 0000, while in binary notation it is 1010; and $353_{(10)}$ in BCD is 0011 0101 0011, while in binary notation it is 101100001. In order to avoid confusing the BCD format with pure binary code, a space is left between each 4-bit group.

The BCD code is comparatively easy to use but is less efficient than the pure binary code. This is because in dealing with higher range decimal numbers more bits are required to make up a BCD word, and this requires more digital circuitry to be associated with data words. In addition, arithmetic operations with BCD numbers are more time-consuming and complex; for example, special adder circuits are required. However, improved interface communication is possible in many applications, and so the use of BCD does help to outweigh some of the disadvantages.

**Excess 3 Code (XS3)**

This code is non-weighted and is a modified form of BCD, and as the name implies each 4-bit number is 3 larger than in BCD. Thus, from Table 2.11 decimal number 5 in XS3 is 1000, while in BCD

**Table 2.11** XS3 code

| Decimal | BCD | XS3 |
|---------|------|------|
| 0 | 0000 | 0011 |
| 1 | 0001 | 0100 |
| 2 | 0010 | 0101 |
| 3 | 0011 | 0110 |
| 4 | 0100 | 0111 |
| 5 | 0101 | 1000 |
| 6 | 0110 | 1001 |
| 7 | 0111 | 1010 |
| 8 | 1000 | 1011 |
| 9 | 1001 | 1100 |

this 4-bit number is decimal 8. The code was developed primarily to simplify arithmetic computations with BCD numbers. As in the case of the BCD code, the XS3 code for all other decimal numbers is obtained by substituting the appropriate 4-bit groups; for example, $25_{(10)} = 0101\ 1000$ and $3471_{(10)} = 0110\ 0111\ 1010\ 0100$. To convert from XS3 to decimal, the decimal equivalent of each 4-bit group is written down and then 3 is subtracted from the number.

### 5–4–2–1 and 7–4–2–1 Codes

These are also 4-bit BCD codes, and are position weighted as indicated by their names (see Table 2.12). As with the standard BCD code, decimal values can be easily obtained by adding the position values for those positions having a binary 1.

**Table 2.12** 5–4–2–1 and 7–4–2–1 codes

| Decimal | 5–4–2–1 | 7–4–2–1 |
|---|---|---|
| 0 | 0000 | 0000 |
| 1 | 0001 | 0001 |
| 2 | 0010 | 0010 |
| 3 | 0011 | 0011 |
| 4 | 0100 | 0100 |
| 5 | 1000 | 0101 |
| 6 | 1001 | 0110 |
| 7 | 1010 | 1000 |
| 8 | 1011 | 1001 |
| 9 | 1100 | 1010 |

### Five-Bit Codes

Although only 4 bits are needed to encode all digits from 0 to 9, the use of more bits usually allows some extra feature in the code, making it easier to operate with, and to provide easier error detection and/or correction features (see later in this chapter). Three examples of 5-bit codes are given in Table 2.13. The

**Table 2.13** Examples of 5-bit codes

| Decimal | 2-out of-5 | 51111 | Shift-counter |
|---|---|---|---|
| 0 | 00011 | 00000 | 00000 |
| 1 | 00101 | 00001 | 00001 |
| 2 | 00110 | 00011 | 00011 |
| 3 | 01001 | 00111 | 00111 |
| 4 | 01010 | 01111 | 01111 |
| 5 | 01100 | 10000 | 11111 |
| 6 | 10001 | 11000 | 11110 |
| 7 | 10010 | 11100 | 11100 |
| 8 | 10100 | 11110 | 11000 |
| 9 | 11000 | 11111 | 10000 |

2-out-of-5 code is unweighted and as will be noted there are only two 1s in each 5-bit number; hence the derivation of its name. In avionic systems, a typical application of this code is associated with frequency selection in a navigation receiver, and generation of logic signals for distance measuring.

The 51111 code is a weighted code, and from its 'pattern' it will be noted that from decimal 1 to 4 a bit 1 is inserted from the right on each successive count, and that from decimal 5 to 9 a bit 1 is inserted from the left. This code pattern facilitates logic circuit operation.

The shift-counter or Johnson code is non-weighted, and like the 51111 code, its pattern of inserting or (shifting) bit 1s and 0s also facilitates logic circuit operation.

**Biquinary and Ring-Counter Codes**

The biquinary (meaning 2 and 5) is a weighted code of 7 bits with only two bits being 1 in each code group, as shown in Table 2.14.

**Table 2.14** Biquinary and ring-counter codes

| Decimal | Biquinary 50 43210 | Ring-counter 9876543210 |
|---|---|---|
| 0 | 01 00001 | 0000000001 |
| 1 | 01 00010 | 0000000010 |
| 2 | 01 00100 | 0000000100 |
| 3 | 01 01000 | 0000001000 |
| 4 | 01 10000 | 0000010000 |
| 5 | 10 00001 | 0000100000 |
| 6 | 10 00010 | 0001000000 |
| 7 | 10 00100 | 0010000000 |
| 8 | 10 01000 | 0100000000 |
| 9 | 10 10000 | 1000000000 |

Since one of these bits must always be in the two left positions, and the other in the five right-most positions, the detection of any errors in the code is made easier. The weighting of the code is 50 43210.

The ring-counter code has only a bit 1 in each code group of 10 bits and is weighted as indicated in Table 2.14. A decimal value is obtained by counting over from the right to the bit 1 starting from the right-most position as zero. The code is applied to certain computer counting and decoding circuits, and derives its name from the fact that in such circuits, a bit 1 is moved around in a closed ring-like connection. When the bit 1 gets to the left-most position, the next count brings the bit 1 around to the zero position to repeat the count from decimal 0 to 9.

## Gray Code

This code, which is also known as the cyclic, unit distance or reflective code, is non-weighted and used largely with optical or mechanical shaft position encoders. A particular application to avionics is the encoding altimeter used in conjunction with a transponder system (see Chapter 9). Table 2.15 shows that in a

**Table 2.15** The Gray code

| Decimal | Gray | Binary |
|---|---|---|
| 0 | 0000 | 0000 |
| 1 | 0001 | 0001 |
| 2 | 0011 | 0010 |
| 3 | 0010 | 0011 |
| 4 | 0100 | 0100 |
| 5 | 0111 | 0101 |
| 6 | 0101 | 0110 |
| 7 | 0100 | 0111 |
| 8 | 1100 | 1000 |
| 9 | 1101 | 1001 |
| 10 | 1111 | 1010 |
| 11 | 1110 | 1011 |
| 12 | 1010 | 1100 |
| 13 | 1011 | 1101 |
| 14 | 1001 | 1110 |
| 15 | 1000 | 1111 |
| . | . | . |
| . | . | . |
| . | . | . |

change from one decimal number to another there is only a 1-bit change between any two successive words in Gray code; in binary code this is not so. For example, in the change from 7 (0111) to 8 (1000) in binary, all four bits change, while in Gray code (the change from 7 to 8 (0100 to 1100)) only the first or MSB changes. Timing and speed changes are therefore minimized by use of the Gray code. Its principal disadvantage is that it must generally be converted into pure binary form for arithmetic operations to be performed.

## Alphanumeric Codes

In addition to the codes we have covered, which are for the purpose of representing only decimal numbers, there are several codes in use for representing alphabetic characters as well as numbers; these are called *alphanumeric* codes. The one having the widest application to avionic systems is the ASCII code listed in Table 2.16. The code initials stand for American Standard Code for Information Interchange. It is a 7-bit binary code (in some cases it may have 6 or 8 bits), so that a total of $2^7 = 128$ different states, alphanumeric characters or control functions can be represented. Each 7-bit code

**Table 2.16** American Standard Code for Information Interchange

| | | Column | | | | | | | |
|---|---|---|---|---|---|---|---|---|---|
| | | 0 | 1 | 2 | 3 | 4 | 5 | 6 | 7 |
| | Bits 765 → | 000 | 001 | 010 | 011 | 100 | 101 | 110 | 111 |
| Row | 4321 ↓ | | | | | | | | |
| 0 | 0000 | NUL | DLE | SP | 0 | @ | P | \ | p |
| 1 | 0001 | SOH | DC1 | ! | 1 | A | Q | a | q |
| 2 | 0010 | STX | DC2 | " | 2 | B | R | b | r |
| 3 | 0011 | ETX | DC3 | # | 3 | C | S | c | s |
| 4 | 0100 | EOT | DC4 | $ | 4 | D | T | d | t |
| 5 | 0101 | ENQ | NAK | % | 5 | E | U | e | u |
| 6 | 0110 | ACK | SYN | & | 6 | F | V | f | v |
| 7 | 0111 | BEL | ETB | ' | 7 | G | W | g | w |
| 8 | 1000 | BS | CAN | ( | 8 | H | X | h | x |
| 9 | 1001 | HT | EM | ) | 9 | I | Y | i | y |
| 10 | 1010 | LF | SUB | * | : | J | Z | j | z |
| 11 | 1011 | VT | ESC | + | ; | K | [ | k | { |
| 12 | 1100 | FF | FS | , | < | L | \ | l | ¦ |
| 13 | 1101 | CR | GS | − | = | M | ] | m | } |
| 14 | 1110 | SO | RS | . | > | N | ⌢ | n | ~ |
| 15 | 1111 | SI | US | / | ? | O | — | o | DEL |

Explanation of special control functions in columns 0, 1, 2 and 7:

| | | | |
|---|---|---|---|
| NUL | Null | DLE | Data Link Escape |
| SOH | Start of Heading | DC1 | Device Control 1 |
| STX | Start of Text | DC2 | Device Control 2 |
| ETX | End of Text | DC3 | Device Control 3 |
| EOT | End of Transmission | DC4 | Device Control 4 |
| ENQ | Enquiry | NAK | Negative Acknowledge |
| ACK | Acknowledge | SYN | Synchronous Idle |
| BEL | Bell (audible signal) | ETB | End of Transmission Block |
| BS | Backspace | CAN | Cancel |
| HT | Horizontal Tabulation (punched card skip) | EM | End of Medium |
| LF | Line Feed | SUB | Substitute |
| VT | Vertical Tabulation | ESC | Escape |
| FF | Form Feed | FS | File Separator |
| CR | Carriage Return | GS | Group Separator |
| SO | Shift Out | RS | Record Separator |
| SI | Shift In | US | Unit Separator |
| SP | Space (blank) | DEL | Delete |

word is made up of a 4-bit group and a 3-bit group, as shown in Fig. 2.1, and in the code table they are arranged in rows and columns.

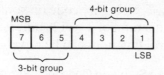

**Fig 2.1** ASCII code word format

To determine the code for a given character, it is first located in the table and then the codes associated with the location column and row are noted. For example, the code for the letter T is required: from the table we note that it is located in column 5, row 4, and also that the most significant 3-bit group is 101, and the least significant 4-bit group is 0100. The code for the letter T is therefore 101 0100.

**Parity in Codes**

The detection of errors in codes and their correction is a very important aspect in the transmission of digital data, and for this purpose a *parity check* method is provided whereby a computer can test whether the total number of bit 1s in a code word is odd or even. For error detection, extra bits called *parity bits* are added to a code word, and this is illustrated in Table 2.17 using the BCD code

**Table 2.17** Introduction of parity bits

| Decimal | BCD | With odd parity bit | With even parity bit |
|---|---|---|---|
| 0 | 0000 | 1 00001 | 0 00000 |
| 1 | 0001 | 0 00010 | 1 00011 |
| 2 | 0010 | 0 00100 | 1 00101 |
| 3 | 0011 | 1 00111 | 0 00110 |
| 4 | 0100 | 0 01000 | 1 01001 |
| 5 | 0101 | 1 01011 | 0 01010 |
| 6 | 0110 | 1 01101 | 0 01100 |
| 7 | 0111 | 0 01110 | 1 01111 |
| 8 | 1000 | 0 10000 | 1 10001 |
| 9 | 1001 | 1 10011 | 0 10010 |

as an example. The bits are added to the right of the standard BCD combination, and establish *even* parity when the total number of bits is an *even* amount, and *odd* parity when there is an *odd* number of 1s in the total number.

In addition to the detection of an error, it is also helpful to be able to correct the error. This is done by adding a *parity word* to the data words, as shown in Table 2.18. The right-most bit of each word is

**Table 2.18** Error correction by using parity word

| | | |
|---|---|---|
| 01111001 ← Parity bit | 01111001 ← | Parity bit |
| 00000111 | 00000111 | |
| 10001111 | 10001111 | |
| 01101110 | 01101110 | Parity error |
| 00110001 | 00111001 | detected |
| 01011000 | 01011000 | |
| 11110001 | 11110001 | |
| 11111000 ← Parity word → | 11111000 | |
| (a) | (b) | |

the odd parity bit, and the last word is an odd parity word, which makes each column of bits have odd parity. When the computer looks at the received data it does a parity check of each word and of each column. Assume, for example, that the data word in row 5 at (a) is received as shown at (b). Since a 1 bit has been inserted in column 4 instead of a 0 bit, the word is now of even parity and so are the bits in column 4 as detected by the parity word. By simply changing the bit to a 0 the computer will correct the data word and store it in memory along with the other six words received, the parity word being no longer required. Bit parity may still be maintained for error detection when using the computer memory.

**Arithmetic Operations**

In many digital systems it is necessary for the basic arithmetic operations of addition, subtraction, multiplication and division to be carried out. Essentially they are the same as in ordinary arithmetic involving decimal numbers, but are obviously restricted to the use of the bits 1 and 0. The rules for all four operations are extensively detailed in the many basic textbooks currently available on the subjects of digital systems, computer technology, etc., and so it is intended here only to review them briefly, and to consider one or two examples of applications.

### Binary Addition

When two 1s are added together their sum is 10, and so in that order of unit, the sum is 0 with a carry of 1. The sum will always be 0 for an even number of 1s (including carries), while for an odd number of 1s the sum will always be 1 including the carries. Table 2.19 illustrates the rules of addition.

**Table 2.19** Rules of binary addition

| A | + | B | Sum | Carry |
|---|---|---|-----|-------|
| 0 |   | 0 | 0   | 0     |
| 1 |   | 0 | 1   | 0     |
| 0 |   | 1 | 1   | 0     |
| 1 |   | 1 | 0   | 1     |

As an example let us consider the addition of 1011 and 1110:

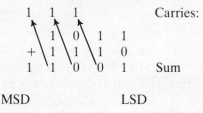

```
              Carries:
   1 1 1
   1 0 1 1
 + 1 1 1 0
 1 1 0 0 1    Sum
MSD       LSD
```

**Binary Subtraction**

When a bit is to be subtracted from a smaller bit, a 1 must be borrowed from the next column to the left. This 1 is a power of 2 higher and hence the subtraction becomes $2 - 1 = 1$. The rules are summarized in Table 2.20. For example, subtract 10101 from 11011:

```
    1 borrowed
      ↗
   11011 = 27
  -10101 = 21
   ─────
   00110 = 6₍₁₀₎
```

$$00110 = 6_{(10)}$$

**Table 2.20** Rules of binary subtraction

| A | − | B | Difference | Borrows |
|---|---|---|------------|---------|
| 0 |   | 0 | 0          | 0       |
| 1 |   | 0 | 1          | 0       |
| 0 |   | 1 | 1          | 1       |
| 1 |   | 1 | 0          | 0       |

The foregoing example is of direct subtraction, but it is more easily carried out electronically on the basis that a binary number has two complements, known as the *ones complement* and the *twos complement*. The former is obtained by changing all the 1s in the number to 0s and all the 0s into 1s. The latter is obtained by adding 1 to the ones complement. Subtraction is achieved by obtaining the ones or twos complement of a number $x$ and then adding it to a number $y$. If the left-hand digit (the sign digit) of the sum is 0, the difference is negative; if it is 1 the difference is positive.

Subtraction of 11101 from 01011 using the ones complement method is as follows:

$$\begin{array}{r} 01011 = 11 \\ -11101 = 29 \end{array} = \begin{array}{r} 01011 \\ +00010 \text{ ones complement} \\ \hline 001101 \end{array}$$

and the difference is therefore $10010 = -18_{(10)}$

Subtraction of 10101 from 11011 using the twos complement method is as follows:

$$\begin{array}{r} 11011 = 27 \\ -10101 = 21 \end{array} = \begin{array}{r} 11011 \\ +01010 \text{ twos complement} \\ \hline 100110 = +6_{(10)} \end{array}$$

### Binary Multiplication

This is easier than in any other system since a multiplier digit can only be a 0 or a 1 to form the partial products which are binary added, and are either zero or exactly the multiplicand. For example, multiply 110101 by 111:

$$\begin{array}{r} 110101 = 53 \\ \times 111 = \phantom{0}7 \\ \hline 110101\phantom{00} \\ 110101\phantom{0} \\ 110101 \\ \hline 101110011 = 371_{(10)} \end{array}$$

In a computer, arithmetic unit multiplication is performed by repeated addition.

### Binary Division

This is also a simple operation since division into a number can only be done once or not at all. No quotient terms other than 1 or 0 are possible. For example, to divide 1001000 ($72_{(10)}$) by 110 ($6_{(10)}$):

$$
\begin{array}{r}
1100 = 12_{(10)} \\
110\overline{)1001000} \\
-110 \phantom{0000} \\
\hline
00110 \phantom{00} \\
-110 \phantom{00} \\
\hline
0000000
\end{array}
$$

Division is performed in a computer arithmetic unit by repeated subtractions.

## Transmission of Digital Data

There are two basic ways in which data are transmitted, processed or otherwise manipulated, and these are designated as serial and parallel (Fig. 2.2).

In serial format the voltage level changes, representing the bits 1 and 0 of a word, occur at a single point in a circuit or on a single line,

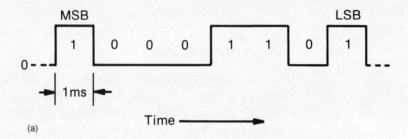

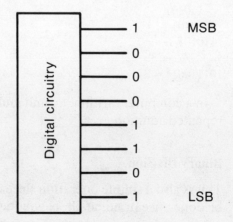

**Fig 2.2** Data transmission for word $10001101_{(2)}$ in (a) serial format, and (b) parallel format

44 Number systems and coding

each bit existing for a specific interval of time; the total transmission time of a word therefore depends on the number of bits. In the example shown, the time taken would be 8 ms. The MSB occurs first since time is considered to be increasing from left to right.

The primary advantage of serial format is that it requires only a single line or channel for transmitting data from one place to another. In addition, since each bit on the single line occurs separately from the other, only one set of digital circuitry is generally needed to process the data. Its disadvantage is the significant amount of time required for the transmission and processing of a serial word.

In the parallel format, all bits of a word are transmitted or processed simultaneously, and as will be noted from Fig. 2.2 a separate line or channel is required for each bit to be transmitted. Since all the bits are available at the same time, digital circuitry must be provided to process each bit, and so this makes processing rather more complex. It does, however, have a clear advantage over the serial format in terms of speed of data transmission.

**Airbus A310**—an example of an aircraft utilizing new electronic display technology. By kind permission of British Caledonian Airways.

# 3 Logic gates and circuits

In avionic systems logical or arithmetic processes are used extensively. They involve complex functions of several variables, the desired functions being realized by switching operations. Much of the design of these systems deals with the interconnection of functional blocks, of complex circuitry, and these in turn consist of basic decision-making elements referred to as *logic gates*, each performing combinational operations on their inputs and so determining the state of their outputs.

Logic gates are of a binary nature, i.e. the inputs and the outputs are in one of two states expressed by the digital notation 1 or 0. Other corresponding expressions are also frequently used, as follows:

1—on; true; high (H); closed; engaged
0—off; false; low (L); open; disengaged

The inherent function of a logic gate, therefore, is equivalent to that of a conventional switch which is a 'two-state' device, and this may be illustrated by considering a simple motor control circuit as shown in Fig. 3.1. In the 'off' position of the control switch the whole

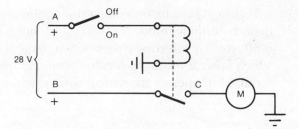

**Fig 3.1** Simple motor control circuit

circuit is 'open', and is in an inactive or logic 0 state. In the 'on' position the switch 'closes' the circuit to the relay coil, causing a positive direct current to pass through the coil; since this voltage is at a 'high' level with respect to ground, the input at A is of a logic 1 state and so the relay is activated. The motor is activated by simultaneous closing of the relay switching contacts, and since the

47

level of the voltage necessary for motor operation is also 'high', the input at B is of logic 1 state. Thus the control switch and the relay perform combined operations on their respective inputs in order to determine a logic 1 state of the output C, and as we shall see later the whole circuit may be considered as a logic function circuit.

The 1 and 0 state designations are arbitrary. For example, if the two states are represented by voltage levels, one may be positive and the other 0 V, one may be negative and the other 0 V, one may be positive and the other negative, both may be positive or both may be negative. Therefore, the application of logic to a system or device may be further defined as follows:

1. positive logic when the more positive potential (high) is consistently selected as the 1 state (as in the circuit of Fig. 3.1, for example);
2. negative logic when the less positive potential (low) is consistently selected as the 1 state;
3. hybrid or mixed logic when both positive or both negative logic is used.

The definitions of (1) and (2) are graphically shown in Fig. 3.2.

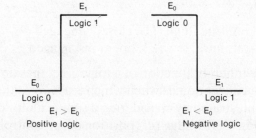

**Fig 3.2** Positive and negative logic

Figure 3.3 is a further illustration of logic gate switching. It, too, depicts a simple motor control circuit, but now control of the circuit is effected by selecting either of two parallel-connected switches which are, for example, located remotely from each other. If it is required to operate the motor from, say, the switch 1 location, the

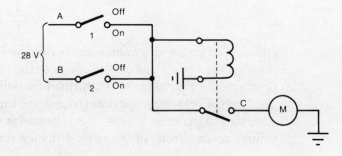

**Fig 3.3** Logic switching

circuit from input A to the relay coil is closed by putting the switch to the 'on' position, thereby producing an active logic 1 state in the coil circuit and at the output C. Switch 2 remains open so that the circuit from input B is in the logic 0 state. The converse would be true were the motor to be operated from the switch 2 location and with switch 1 'off'.

The primary difference to note between this switching arrangement and the one shown in Fig. 3.1 is that a logic 1 state at output C can be produced when *either* switch 1 or switch 2 is closed, *or both* are closed. This circuit also has an equivalent in logic circuit terms which we shall come to later.

**Gates and their Symbols**

There are several types of gates applied to the circuits of the various avionic systems in current use, and regardless of their circuit design details they all act as though they consisted of electrically controlled switches, opening and closing their appropriate signal flow lines at speeds far beyond the realms attainable by their conventional counterparts.

The three basic gates perform the functions AND, OR and NOT (or as it is more usually termed, an 'inverting' function). All logic circuits are simply combinations of these gates and their functions, and because the circuit output can always be predicted from the combination of inputs which is present, they are, in the broadest sense, designated as *combinational logic circuits*. In each case the switching is performed by either junction diodes, transistors or by a combination of both, such combinations giving rise to specific circuits classified as diode–transistor logic (DTL) and transistor–transistor logic (TTL or $T^2L$).

### The AND Gate

The circuit of a positive logic gate using junction diodes and performing the AND function is shown in Fig. 3.4. In this circuit, a 'high' or logic 1 output state is only produced when the inputs at A *and* B are 'high'; when either input is 'low' or logic 0, or both inputs are 'low', then the output state at X is also 'low'. This may be seen by considering the waveforms when certain numeric values, or logic levels, of voltage are applied to the inputs; in this case 0 V represents a logic 1 state, and $-6$ V represents a logic 0 state.

During the period $T_0-T_1$ a logic 0 is applied to both inputs A and B and as a bias to the cathodes of $D_1$ and $D_2$. Since the inputs are negative with respect to the voltage at the anodes, both diodes will be forward-biased and they will conduct, i.e., 'switches closed'.

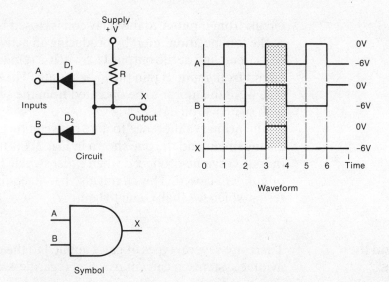

**Fig 3.4** AND gate

The function of the load resistor $R$ is to develop a voltage drop which separates the diode supply voltage from the output of the gate when the diodes are conducting. Therefore, under the input logic conditions assumed, the state at the output X of the gate will also be logic 0.

From time $T_1$ to $T_2$, a logic 1 is applied to input A while input B remains at logic 0. Diode $D_2$ therefore remains forward-biased and conducting, and it clamps the anode of $D_1$ to $-6$ V so that it is now reverse-biased ('switch open') and will not conduct. Because of the voltage-dropping function of resistor $R$ while $D_2$ is conducting, the output state at X is also clamped at logic 0. During the period $T_2-T_3$ the logic states at A and B are the reverse of those just described, and they also produce a logic 0 at output X.

The period $T_3-T_4$ represents the fourth and last combination of logic states which is possible for the basic 2-input AND gate shown in Fig. 3.4, and as emphasized in the waveform diagram it is the only combination which will produce an output state of logic 1. With 0 V (logic 1) at inputs A and B, the anodes of $D_1$ and $D_2$ are positive with respect to their cathodes (forward-biased), and so they will conduct and clamp the output state at X to logic 1.

In summarizing the foregoing, we can therefore say that an AND gate is an 'all or nothing' gate.

*Symbols*

In aircraft systems which are based on digital logic, the circuit arrangements may involve the use of only a few logic gates, or literally thousands of them depending on the overall function to be performed by a particular system. Although at this stage we have only studied one type of gate, it should be apparent that if a logic circuit diagram is drawn for a complete system, and if it also shows the internal circuit of every gate used, the diagram could be a very complex one to trace through. In order, therefore, to simplify diagrams as much as possible, the internal circuit arrangements are omitted, and the gates are represented by corresponding distinctively shaped symbols which conform to accepted standards.

The symbol for the AND gate we have so far studied is shown in Fig. 3.4. The power supply and ground connections essential for gate operation are always taken for granted, and so to further simplify circuit diagrams they are not shown on gate symbols.

Variations in the symbol shapes adopted will be found in some literature, but those in this book are adopted in the majority of manuals related to aircraft systems.

There are several other aspects related to the use of symbols, but as these are concerned with the overall interpretation of practical circuit diagrams, they will be described at a later appropriate stage.

*Truth Tables*

Truth tables are mathematical tables which display all possible combinations of input logic states and their corresponding output states in terms of the 'bits' 0 and 1. The tables are important tools in the design of all logic circuits in that they describe logical statements in a concise manner. Having already dealt with one type of circuit we can now briefly study the construction of its table as a preliminary to those of the other circuits yet to be described.

As will be noted from Fig. 3.5, a table is a rectangular coordinate presentation, the columns representing the inputs and outputs, and the rows representing the logic combinations. The number of different possible combinations is expressed by $2^n$, where $n$ is the number of inputs. Thus, for a basic 2-input gate the possible combinations are $2^n = 2^2 = 4$, and so its truth table has two input

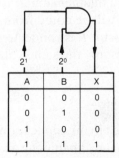

**Fig 3.5** Truth table of an AND gate

| $2^1$ | $2^0$ | |
|---|---|---|
| A | B | X |
| 0 | 0 | 0 |
| 0 | 1 | 0 |
| 1 | 0 | 0 |
| 1 | 1 | 1 |

Microelectronics in Aircraft Systems 51

columns and four rows. For a 3-input gate there would be eight possible combinations, and so on.

Examples of multiple-input logic gates will be covered later in this chapter, but at this point it is of interest to note how the sequencing of the 0s and the 1s which make up the logic combinations is determined. This sequencing is of importance in order to identify every possible output of a gate.

The method adopted is, quite simply, to identify all the inputs with 2s which are raised to a power based upon the input positions in a table. Let us refer once again to Fig. 3.5; the table has two input columns, and working from right to left (this sequence always applies) we identify column B with $2^0$ and column A with $2^1$. Since 2 raised to zero power equals 1, one 0 and one 1 are alternately placed in each row of the column. Two raised to the power of 1 equals 2; therefore, two 0s and two 1s are alternately placed in each row of the column identified by $2^1$.

Some truth tables may indicate logic combinations by the level designation H and L, or by the arbitrary voltage values; an example of the latter, based on the values used in describing the AND gate operation, is shown in Fig. 3.6.

| A | B | X |
|---|---|---|
| -6V | -6V | -6V |
| 0V | -6V | -6V |
| -6V | 0V | -6V |
| 0V | 0V | 0V |

**Fig 3.6** Electrical truth table

### The OR Gate

The circuit of a positive logic gate using junction diodes and performing the OR function is shown in Fig. 3.7, together with its symbol and truth table. In this circuit, a 'high' or logic 1 output state is produced when one *or* other of the inputs is 'high', *or* when both are 'high'. When both inputs are 'low' or logic 0, the output state is 'low'. An OR gate is, therefore, an 'any or all' gate.

The circuit operates in a manner similar to that already described for the AND gate; but as will be noted from Fig. 3.7, the diodes are reversed and the load resistor *R* is connected to a negative potential. The waveforms, which assume the same voltage levels as before, are also shown.

### NOT or INVERTING Logic

In a large number of logic circuit applications it is necessary to change the state of a signal by *not* having a voltage at the output

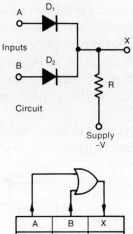
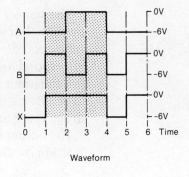

Waveform

| A | B | X |
|---|---|---|
| 0 | 0 | 0 |
| 0 | 1 | 1 |
| 1 | 0 | 1 |
| 1 | 1 | 1 |

**Fig 3.7** OR gate      Symbol and truth table

every time there is a voltage at the input, or vice versa. In other words, the function of such a circuit is to 'invert' the input signal such that the output signal is always of the opposite state.

An example of an inverter circuit using an n-p-n transistor is shown in Fig. 3.8. Resistor $R_1$ is in series with the transistor base and input A, while resistor $R_2$ is connected between the base and the negative voltage supply $V_{BB}$. This ensures that when A is at logic 0 the base–emitter junction is reverse-biased to produce a logic 1 state at output B. When input A is at logic 1 state, the transistor conducts to produce a logic 0 state at the output. If a p-n-p transistor is used for the inverting function, the polarities $V_{CC}$ and $V_{BB}$ are reversed.

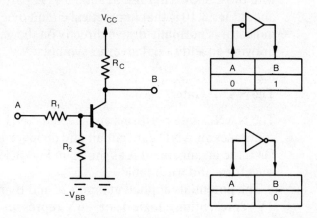

**Fig 3.8** Inverter

Microelectronics in Aircraft Systems

The symbol for an inverter is the same as that adopted for an amplifier but with the addition of a small circle (called a 'state indicator') drawn at either the input or the output side to correspond to the truth tables, as shown in Fig. 3.8. When the circle is at the input, it means that the input signal must be 'low' for it to be an activating signal; when at the output it means that the output of an activated function is 'low'.

The output of an AND or OR gate circuit can become a NOT function by the addition of an inverter, as shown in Fig. 3.9, the

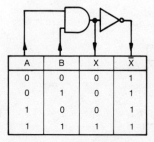
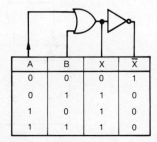

**Fig 3.9** NOT function applied to AND and OR gates

appropriate gates then becoming known respectively as NAND (which is a contraction of NOT AND) and NOR (which is a contraction of NOT OR). In order to emphasize the NOT function, a line is drawn over a letter in a logic expression, as indicated in the last column of the truth table in Fig. 3.9.

*Inhibited Gates*

In many cases the NOT function is used in conjunction with the input to an AND or OR gate; the gate is then said to be *negated* or *inhibited*. Figure 3.10 illustrates inhibited inputs to both types of gate, and the corresponding truth tables; these should be compared with those shown in Figs. 3.5 and 3.7. A particular point to note about Fig. 3.10 is that in practical circuit diagrams the state indicator is normally drawn directly on the gate symbols rather than drawing an additional inverter symbol.

**The NAND Gate**

The NAND gate performs an inverted AND function. Its circuit comprises an AND gate circuit and an inverter circuit, and based on DTL the arrangement is as shown in Fig. 3.11, together with its waveform and truth table.

The potentials applied at inputs A and B are assumed to be either 0 V representing a logic 1, or −6 V representing a logic 0. If a

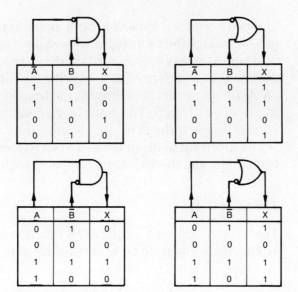

**Fig 3.10** Negated AND and OR gates

logic 0 is applied to either diode, or to both simultaneously, a total of 12 V is established across the voltage divider formed by $R_1$ and $R_2$. Resistor $R_2$ causes 10 V to be dropped across it, and this establishes a potential of $-4$ V at the base of the transistor. The transistor is therefore forward-biased and will conduct. Current flow through the transistor clamps the output to 0 V, i.e., logic 1.

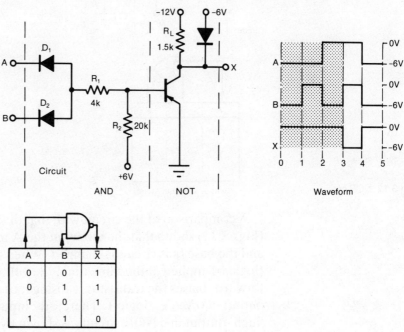

**Fig 3.11** NAND gate

Symbol and truth table

Microelectronics in Aircraft Systems

When a logic 1 is applied to both inputs at the same instant, a total of 6 V is established across the voltage divider. $R_2$ now causes 5 V to be dropped across it, and so establishes a potential of $+1$ V at the transistor base; this reverse-biases the transistor, causing it to cut off. With no current flow through the transistor the output is clamped at $-6$ V, i.e., logic 0. The diode in the collector circuit allows current to flow through $R_L$, thus holding the collector at $-6$ V.

The inversion of the outputs of a NAND gate are shown in its truth table, and this should be compared with that given in Fig. 3.5.

### The NOR Gate

This gate performs an inverted OR function and its circuit arrangement and truth table are illustrated in Fig. 3.12.

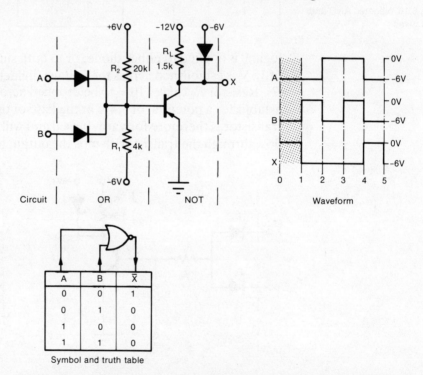

**Fig 3.12** NOR gate

A comparison of the circuit with that of the NAND gate (Fig. 3.11) shows that the diodes at the A and B inputs are reversed and the base bias circuit is changed. When both diodes have $-6$ V (logic 0) applied at the same time, the voltage drop of 2 V across $R_1$ forward-biases the transistor, causing it to conduct and produce an output of 0 V, i.e., logic 1. Therefore, inputs must be 'low' to give a 'high' output in a NOR circuit, as shown by the truth table.

When a logic 1 input is applied to either diode the corresponding diode will conduct so that the voltage drop across $R_1$ will equal 6 V, making the transistor base and emitter at the same potential, thus cutting off the transistor. The cut-off condition clamps the output at −6 V as established by the output diode. Any 'high' input, i.e., at A or B or both, therefore causes a 'low' output.

### Exclusive OR and NOR Gates

The exclusive OR gate is basically a combination of two inhibited AND gates and an OR gate, and as will be noted from Fig. 3.13(a) it develops an output pulse when either input A or B is present, but not when both are present. Its principal applications are in binary adder circuits and parity checking circuits.

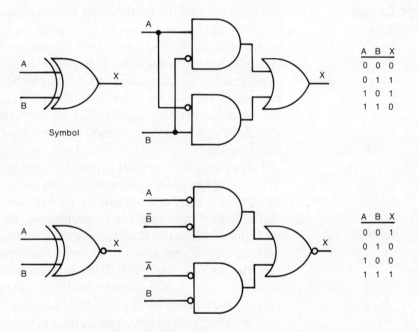

**Fig 3.13** (a) Exclusive OR gate ($X = \overline{A}B + A\overline{B}$).
(b) Exclusive NOR gate ($X = \overline{AB} + AB$)

The exclusive NOR gate shown in Fig. 3.13(b) is sometimes referred to as an equivalence circuit or a binary comparator, and from its truth table it will be noted that its output is the complement of the exclusive OR gate.

### Wired Gates

In some cases inputs may be connected together in the configurations known as 'wired OR' and 'wired AND', and their symbols are shown in Fig. 3.14.

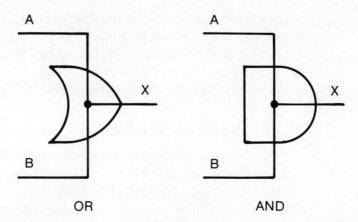

**Fig 3.14** Wired gates          OR          AND

**Logic Circuit Equations**

We have studied the relationship between input and output states based on the binary number system, and we have seen how truth tables are compiled to reveal such relationship for any logic gate combination. Although the truth tables do provide a concise means of presenting logical statements, these can be further simplified by expressing them in the form of algebraic equations. These are known as Boolean equations after the mathematician George Boole who developed a system of algebra solely for expressing binary logic relationships.

This system is somewhat similar to conventional algebra. Whereas in the latter system letters are used in equations to signify numerical values, and are related by symbols signifying the mathematical operations to be performed, such as addition or multiplication, in the Boolean system the letters signify *statements*, and the symbols relating them signify the *logical operation* or *function* such as 'and', 'or'. Thus, by remembering the fact that the logic values of variables are restricted to the two possible 'truth values' of a statement, regardless of whether they are expressed as true or false, high or low, 1 or 0, then in conjunction with the truth tables, or even from a logic circuit diagram, the formulation of a Boolean equation is a much simpler process.

The symbols adopted in equations for signifying logical operations, and these are only in terms of 'product of' and 'sum of', are the same as those adopted for the equivalent mathematical operations, namely × or •, and +, but as logical statements are *not* mathematical the symbols are *not used* in the same way. The letters adopted in an equation signify the logical statements or, in terms of practical circuitry, they signify the signal conditions at the inputs and the outputs of gates either singly or in combination. Table 3.1 shows how the equations are developed for the three basic logic

**Table 3.1** Derivation of logic circuit equations

| Gate | Logical statement | Equation |
|---|---|---|
| AND gate (A, B → X) | If, and only if, inputs $A$ and $B$ are logic 1 at the same time, there will be a logic 1 output at $X$. | $A \cdot B = X$ may also be written $AB = X$ |
| OR gate (A, B → X) | If any one, or all, of inputs are logic 1, there will be a logic 1 output at $X$. | $A + B = X$ |
| NOT gate (A → X) | If input $A$ is either logic 1 or logic 0, the output at $X$ will not be $A$. | $\overline{A} = X$ |

gates and functions, and from this it should be particularly noted how the 'product' and 'sum of' symbols are applied to the AND and OR functions. Some examples of the use of equations in relation to circuits containing combinations of gates will be given in Chapter 7.

**Karnaugh Maps**

These are used as an alternative to Boolean logic equations. They minimize logic expressions in that a logical function can be displayed diagrammatically on a set of squares, the number of which is derived from the expression used for compiling a truth table, namely $2^n$. The relationship between a map and the equation $X = A\overline{B} + \overline{A}B$ (the exclusive OR function) is shown in Fig. 3.15 and is intended only as an example of how a map is constructed.

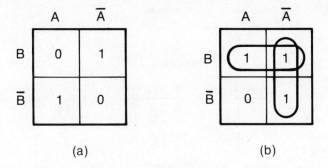

**Fig 3.15** Karnaugh map:
(a) $X = A\overline{B} + \overline{A}B$.
(b) $X = AB + \overline{A}B + \overline{A}\overline{B}$

For this equation and from the above expression, the number of squares needed is four, and they are labelled in rows and columns such that each square represents a different combination of the two variables—in this case $AB$, $\overline{AB}$, $A\overline{B}$ and $\overline{A}B$. Each square contains a plot of the equation in terms of binary 1 and binary 0. To simplify an equation adjacent squares containing a binary 1 are looped

together, and this indicates that the corresponding terms in the equation being mapped have been combined, and any terms of the form $A\overline{A}$ have been eliminated. Squares are only looped together in multiples of two. In considering the equation $X = AB + \overline{A}B + A\overline{B}$ shown in Fig. 3.15, the equation simplifies to
$X = \overline{A}(B + \overline{B}) + B(A + \overline{A}) = \overline{A} + B$.

## Dual Functions of Logic Gates

Our explanation of the operation of logic gates has so far assumed the use of positive logic level assignments. If, however, the original voltage levels considered for explanatory purposes are now translated into binary data on the assumption of negative logic, it will be found that the appropriate truth tables, when compiled, indicate that the gates are performing a different function. For example, let us consider once again the AND gate shown in Fig. 3.4. With negative logic assigned the 0 V level will now be logic 0 and the $-6$ V level will be logic 1. Thus, in compiling the truth table it will be found that, apart from the input sequencing not being the same, the gate function has changed from AND to OR (see Fig. 3.16). With negative logic assigned to an OR gate it will be similarly found that it changes its function to that of an AND gate, and this is also shown in Fig. 3.16. A knowledge of the logic levels

| Positive logic | | | | Negative logic | | |
|---|---|---|---|---|---|---|
| A | B | X | | A | B | X |
| 0 | 0 | 0 | | 1 | 1 | 1 |
| 0 | 1 | 0 | | 1 | 0 | 1 |
| 1 | 0 | 0 | | 0 | 1 | 1 |
| 1 | 1 | 1 | | 0 | 0 | 0 |
| AND | | | = | OR | | |

| Positive logic | | | | Negative logic | | |
|---|---|---|---|---|---|---|
| A | B | X | | A | B | X |
| 0 | 0 | 0 | | 1 | 1 | 1 |
| 0 | 1 | 1 | | 1 | 0 | 0 |
| 1 | 0 | 1 | | 0 | 1 | 0 |
| 1 | 1 | 1 | | 0 | 0 | 0 |
| OR | | | = | AND | | |

**Fig 3.16** Dual functions of gates

assigned to digital circuits is of importance in circuit tracing and when troubleshooting, and due attention should be paid to operating notes and acceptable voltage values given on appropriate diagrams.

NAND and NOR gates can also perform respectively the AND

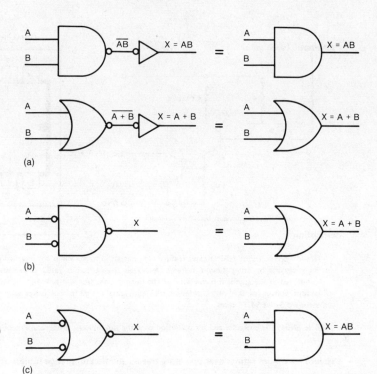

**Fig 3.17** Dual functions of NAND and NOR gates. (a) AND and OR functions. (b) OR function with a NAND gate. (c) AND function with a NOR gate

and OR functions, but in such cases this is accomplished by additional inversion (the effect of one inversion is always cancelled by adding a second) at the output as shown in Fig. 3.17(a). It is also possible for a NAND gate to perform an OR function and a NOR gate to perform an AND function, and as will be noted from Figs 3.17(b) and (c) this is done by inverting the inputs as well as the outputs. Thus, any of the three basic logic functions can be performed with either a NAND gate or a NOR gate, permitting some economy to be achieved in applying them to certain digital circuits.

**Fabrication of Gates**

Gates are fabricated as IC packs either in dual, triple or quadruple circuit arrangements. Figure 3.18 illustrates a typical presentation of manufacturer's operating data, which in this example relates to a quadruple 2-input NAND circuit arrangement contained within a dual-in-line (DIL) pack monolithic IC. The numbered squares represent the connecting pins.

**Logic Concepts and 'Non-Electronic' Systems**

In what may be termed the more conventional type of electrical system, it is necessary to perform switching operations either through the medium of toggle switches or relay contacts.

schematic (each gate)

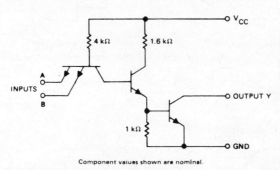

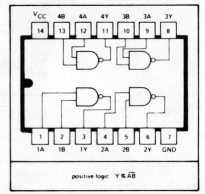

Component values shown are nominal.

### description

These open-collector NAND gates feature high output voltage ratings for interfacing with low-threshold-voltage MOS logic circuits or other 12-volt systems. Although the output is rated to withstand 15 volts, the $V_{CC}$ terminal is connected to the standard 5-volt source. The output transistor will sink 16 milliamperes while maintaining a low-level output voltage of 0.4 volt maximum thus providing a high-fan-out driver with the nominal power dissipation of standard Series 54/74 gates.

The SN5426 is characterized for operation over the full military temperature range of $-55°C$ to $125°C$.

### absolute maximum ratings over operating free-air temperature range (unless otherwise noted)

Supply voltage $V_{CC}$ (see Note 1) . . . . . . . . . . . . . . . . . . . . . . . . 7 V
Input voltage (see Note 1) . . . . . . . . . . . . . . . . . . . . . . . . . . . 5.5 V
Output voltage (see Notes 1 and 2) . . . . . . . . . . . . . . . . . . . . . . . 15 V
Operating free-air temperature range: SN5426 Circuits . . . . . . . . . . . . $-55°C$ to $125°C$
Storage temperature range . . . . . . . . . . . . . . . . . . . . . . . . . $-65°C$ to $150°C$

NOTES: 1. Voltage values are with respect to network ground terminal.
2. This is the maximum voltage which should be applied to any output when it is in the off state.

### recommended operating conditions

|  | SN5426 | | | | | | UNIT |
| --- | --- | --- | --- | --- | --- | --- | --- |
|  | MIN | NOM | MAX | MIN | NOM | MAX |  |
| Supply voltage | 4.5 | 5 | 5.5 |  |  |  | V |
| Output voltage, $V_{OH}$ |  |  | 15 |  |  |  | V |
| Low-level output current, $I_{OL}$ |  |  | 16 |  |  |  | mA |
| Operating free-air temperature range, $T_A$ | -55 | 25 | 125 |  |  |  | °C |

**Fig 3.18** Logic gate data

Comparable operations are also performed in mechanical systems, for example, the opening and closing of valves. It is therefore possible to interpret the operations in terms of logic gate functions, and once an understanding of these functions, and their equivalent symbols, has been gained the reader will find that such understanding may be further developed by studying any chosen electrical circuit diagram or functional diagram of a mechanical system, and then depicting all, or parts, of the diagram in logic form.

As a guide to such exercises, let us refer once again to the simple motor control circuit shown in Fig. 3.1. The control switch and the relay correspond to switching diodes or transistors, and because closing of the switch and the relay contacts is the only combination that will produce the logic 1 state required to activate the motor, the switch and relay may be considered as a logic gate performing an AND function. The circuit and signal functions can therefore be depicted simply as shown in Fig. 3.19(a).

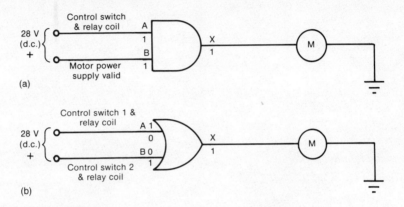

**Fig 3.19** Logic functions of a motor control circuit

The motor control circuit shown in Fig. 3.8 may be similarly considered. In this case the logic 1 state required for motor operation can be effected by selecting either one of two parallel-connected switches in order to enable, i.e., energize the relay. Such a circuit therefore has an equivalent OR function and can be depicted as in Fig. 3.19(b).

Figure 3.20 depicts a hydraulic system in a very elementary form.

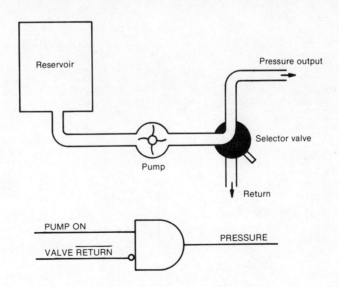

**Fig 3.20** Logic function of a simple hydraulic system

Microelectronics in Aircraft Systems 63

In order to obtain a pressure output the pump must be running, and the selector valve must NOT be in the 'return' position. Interpreting the diagram as a logic gate, we can therefore state that it performs an inhibited AND function.

# 4 Logic devices

Since logic gates are fabricated as IC packs they can be used as individual devices in digital circuits to perform switching functions. It is, however, necessary for them to form parts of numerous other individual 'building-block' devices which are designed to perform such other functions as short-term data storage, data shifting, counting, multiplexing, etc. The purposes and operating fundamentals of some examples of these devices are briefly reviewed in this chapter.

**Flip–Flops**

A flip–flop is a bistable multivibrator circuit within the sequential digital logic family of devices, and has the basic function of storing a single bit of binary data. It is so called because the application of a suitable pulse at one input causes it to 'flip' into one of its two stable states and remain latched in that state, until a pulse at a second input causes it to 'flop' into the other state. The device has two output terminals normally designated $Q$ and $\overline{Q}$, and the logic states of these terminals are referred to respectively as the *normal* or *true*, and the *complement*. When the normal state is logic 1 and the complement is logic 0, the device is said to be *set*; and conversely it is said to be *reset* when the normal state is logic 0 and the complement is logic 1. For such applications as counters and shift registers, flip–flops are operated in synchronism with a *clocked* or *strobed* pulse train, derived from an astable or free-running multivibrator, or a crystal oscillator, and applied to a third input. They consist of NAND or NOR gates and are fabricated as digital ICs. The basic types of flip–flops, their symbols and truth tables are shown in Table 4.1. The term 'indeterminate' noted in some cases means that for the inputs shown the flip–flops may switch to reverse the output states or they may remain in an existing condition; in other words, they get into a state of 'limbo' and so produce undesirable operation.

The fundamental operation of flip–flops may be understood from Fig. 4.1, which shows the arrangement of the basic S–R (set–reset)

**Table 4.1** Basic flip–flops

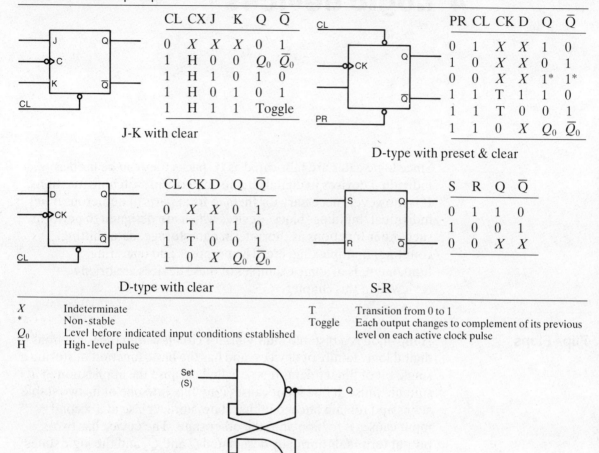

J-K with clear

| CL | CX | J | K | Q | $\bar{Q}$ |
|----|----|---|---|---|-----|
| 0 | X | X | X | 0 | 1 |
| 1 | H | 0 | 0 | $Q_0$ | $\bar{Q}_0$ |
| 1 | H | 1 | 0 | 1 | 0 |
| 1 | H | 0 | 1 | 0 | 1 |
| 1 | H | 1 | 1 | Toggle | |

D-type with preset & clear

| PR | CL | CK | D | Q | $\bar{Q}$ |
|----|----|----|---|---|-----|
| 0 | 1 | X | X | 1 | 0 |
| 1 | 0 | X | X | 0 | 1 |
| 0 | 0 | X | X | 1* | 1* |
| 1 | 1 | T | 1 | 1 | 0 |
| 1 | 1 | T | 0 | 0 | 1 |
| 1 | 1 | 0 | X | $Q_0$ | $\bar{Q}_0$ |

D-type with clear

| CL | CK | D | Q | $\bar{Q}$ |
|----|----|---|---|-----|
| 0 | X | X | 0 | 1 |
| 1 | T | 1 | 1 | 0 |
| 1 | T | 0 | 0 | 1 |
| 1 | 0 | X | $Q_0$ | $\bar{Q}_0$ |

S-R

| S | R | Q | $\bar{Q}$ |
|---|---|---|-----|
| 0 | 1 | 1 | 0 |
| 1 | 0 | 0 | 1 |
| 0 | 0 | X | X |

| | | |
|---|---|---|
| X | Indeterminate | |
| * | Non-stable | |
| $Q_0$ | Level before indicated input conditions established | |
| H | High-level pulse | |
| T | Transition from 0 to 1 | |
| Toggle | Each output changes to complement of its previous level on each active clock pulse | |

**Fig 4.1** S–R flip–flop or latch

type using NAND gates. Let us assume for the moment that power is applied, the R and S inputs are omitted, and that the outputs of gates 1 and 2 are logic 0 and logic 1 respectively. Both gates will remain latched in these two logic states by virtue of their cross-coupled outputs. Assume now that R and S are connected and that they are logic 1 and logic 0 respectively; then both gates will flip to the states whereby by the normal output $Q$ is logic 1 and $\bar{Q}$ is logic 0.

In some circuits flip–flops may be used in what is termed a *master–slave* arrangement, as shown in Fig. 4.2. The purpose of this is to enable flip–flops to slave their input signal before they change their output signal. The circuit in this example is made up of two identical and clocked S–R flip–flops which are simply gated latches, but with two data inputs rather than one, and an inverter. The clock

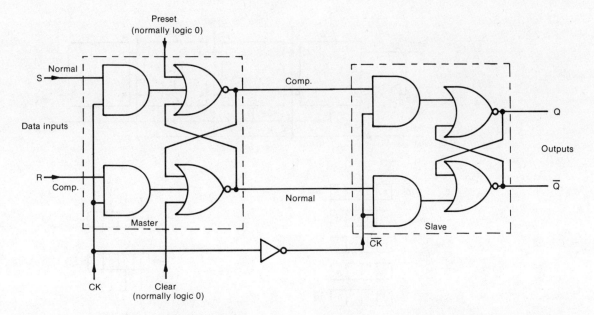

**Fig 4.2** Master–slave flip–flop

signal ($\overline{CK}$) to the slave provides the required store-before-charge action. When the main clock ($CK$) changes from logic 0 to logic 1, the slave holds one bit while the master accepts a new bit. When $CK$ goes to logic 0 again, the master holds the new bit while the slave releases the old one and begins transmitting the new information straight through from the master. Thus, when this type of flip–flop is used with others for, say, shifting data, all the flip–flops get ready to shift when the main clock $CK$ goes logic 1, and they complete the shift when it returns to logic 0.

The 'preset' and 'clear' inputs are normally kept at logic 0 and have no effect on the operation. Putting a momentary logic 1 pulse into the preset input while $CK$ is logic 0 (between clock pulses) will store a binary 1 in the flip–flop. Similarly, a momentary logic 1 pulse into the clear input between clock pulses will store a binary 0. These input features are used in some shift registers where old data need clearing out to leave only binary 0s stored, or to preset the register to all binary 1s. Another use is the loading of parallel data into a register.

**Shift Registers**

A shift register is one of the most versatile of all sequential logic circuits. It is made up of binary storage elements which are generally J–K flip–flops cascaded in such a way that the bits stored can be moved or shifted in serial form from one element to an adjacent element, by the application of a single input clock pulse. Because of this ability it can perform a wide variety of logic operations.

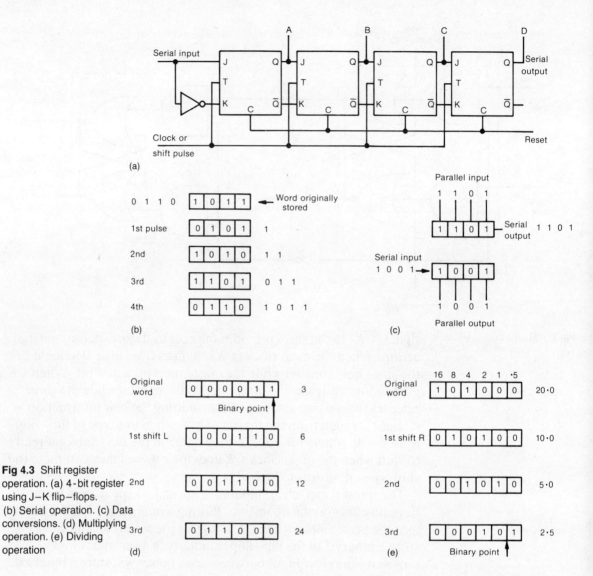

**Fig 4.3** Shift register operation. (a) 4-bit register using J-K flip-flops. (b) Serial operation. (c) Data conversions. (d) Multiplying operation. (e) Dividing operation

Figure 4.3 shows the fundamental operation of a register consisting of four storage elements.

Assume that the binary word 1011 is currently stored in the register and that another word 0110 is generated externally as a serial input. After one clock pulse, the number stored initially is shifted one bit position to the right so that the right-most bit is then lost. At the same time, the first bit of the serial input word is shifted into the left-most bit position of the register. This shifting and 'losing' of stored bits occurs as a result of the flip–flops being set and reset, and as will be seen from diagram (b) after the fourth clock pulse has occurred the serial input word 0110 has been shifted in and is now stored in the register.

68  Logic devices

Shift registers can also be provided with both parallel inputs and outputs, and they also have the ability to perform serial-to-parallel and parallel-to-serial data conversions. The method of conversion is also shown in Fig. 4.3. These devices can also perform such arithmetic operations as multiplication and division, as shown in Fig. 4.3(d) and (e). Shifting a stored word to the left has the effect of multiplying by some power of 2, while shifting to the right has the effect of dividing by some power of 2. The multiplying factor and the dividing ratio is $2^n$, where $n$ is the number of shift left and shift right operations respectively.

In addition to the foregoing applications, shift registers may be used as short-term memories and as sequencers or ring counters.

Registers using bipolar logic circuits and implemented with J–K flip–flops are limited in bit capacity, and so where more storage is needed registers of the MOS IC type are used. There are two types of MOS registers, the static in which clock pulses may be stopped without data being lost, and the dynamic in which data will be lost as a result of stopping clock pulses. The use of either one is largely a matter of 'trading-off' such factors as power consumption and complexity; e.g., a static register consumes more power and is more complex, while a dynamic register has lower power dissipation, operates at higher speeds and is simpler.

**Clocks**

As already mentioned in connection with shift registers, their operation (as indeed of most sequential logic circuit devices) is dependent on the input of pulses that will step them through their operating stages. The pulses are generated by circuits known as *clock oscillators*, the most common of which are some form of free-running multivibrator, or by a *one-shot* (one input, one output) or monostable multivibrator. Such circuits can be constructed with discrete components or with logic gates. A clock oscillator generates rectangular output pulses with a specific frequency, duty cycle and amplitude, and the one-shot generates a rectangular output pulse of a specific time duration each time it receives an input trigger pulse.

**Counters**

Counters are also sequential logic circuit devices made up of flip–flops and used for counting the number of pulses applied to them. In relation to aircraft systems they are in many cases used for the digital display of data, and their operating fundamentals will therefore be covered in Chapter 5.

**Encoders and Decoders**

These are combinational logic circuits, the former being one that accepts one or more inputs and generates a multibit binary output code word, while the latter detects the presence of a specific binary coded word and provides a corresponding binary signal output. Examples of both circuits and their operating fundamentals will also be covered in Chapter 5.

**Multiplexers**

The circuits of many avionic systems are required to select and handle input signals from a number of different sources, and to transfer such data to a common single output without any loss of identity of an individual signal. The data and selection process is referred to as *multiplexing*, and a combinational logic circuit performing this function is known as a data selector, or more usually as a *multiplexer* (abbreviated as MUX or MPX).

There are two basic types of circuit—analog and digital. For analog applications relays, and bipolar or MOSFET switches, are widely used. For digital applications, a MUX can be simply constructed with standard logic gates; this arrangement also permits an analog application.

A very basic form of multiplexer logic is shown in Fig. 4.4, which serves generally to illustrate the fundamental operating principle

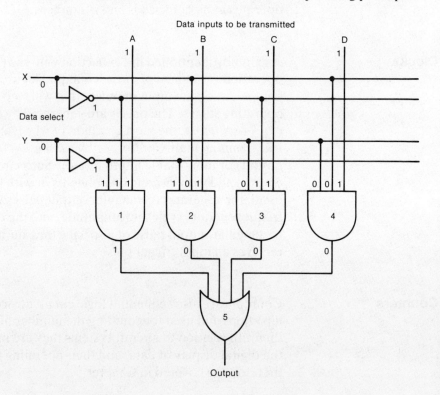

**Fig 4.4** Multiplexer

70 Logic devices

involved. It has four AND gates to which four data inputs to be transmitted are connected, and an OR gate which provides the single output connection to the line (or data highway, as it is known) along which transmission takes place. A data input is selected by the logic states existing at points X and Y and at inverter outputs. The data inputs are each logic 1, and if $X$ and $Y$ are logic 0, then $X$ will apply this as an input to gates 2 and 4, and $Y$ will apply logic 0 to gates 1 and 3. Since the inputs from both $X$ and $Y$ are inverted, logic 1 will be simultaneously applied to all four NAND gates. Gate 1 is the only one with three logic 1 inputs and so it will enable data input $A$ to be transmitted to the highway via the OR gate 5. From the table given in Fig. 3.11, the reader should have no difficulty in working out the logic states which will enable data inputs $B$, $C$ and $D$ to be successively transmitted.

The basic operating modes of a multiplexer may be either *time division multiplexing* (TDM) or *frequency division multiplexing* (FDM). With TDM each of the input signals is sampled and transmitted sequentially so that each signal is transmitted through the common output during predetermined time intervals. The basic principle is shown in Fig. 4.5. If four data channels are operating at 300 bits/s, then the duration of one bit is 1/300 or 3·3 ms; and if the data word is 7 bits, the word occupies a time slot of 23·31 ms. The data present at the inputs of channels 1, 2, 3 and 4 are fed into the appropriate buffer stores and held there until the store is given access to the common output line or data highway. The clock pulses are applied to each gate in turn to sample the stored information and apply it to the data highway for a time period equal to the duration of the word, i.e., 23·31 ms. For example, when clock pulse 1 enables store A for 23·31 ms, the stored data is transmitted to the data highway, and since in TDM the bit rate on the highway is the sum of the individual channel bit rates, then the store A data will be at the rate of 1200 bits/s. Clock pulse 1 then ends and store A goes off-line. Store B is now enabled by a clock pulse to allow one character of the data to be sent to the highway, and after 23·31 ms store B goes off-line and store C is enabled; and so on until all four channels have been connected to the highway. This cycle is repeated to allow other characters in each store to be transmitted. The gates at the receiving end of the highway convert the individual bit rates back to those at which channels are operating; in this case 300 bits/s.

In FDM, transmitted data signals contain several carrier waves each of a different frequency, and separately modulated with a different input signal.

Besides providing a convenient means of selecting one of several

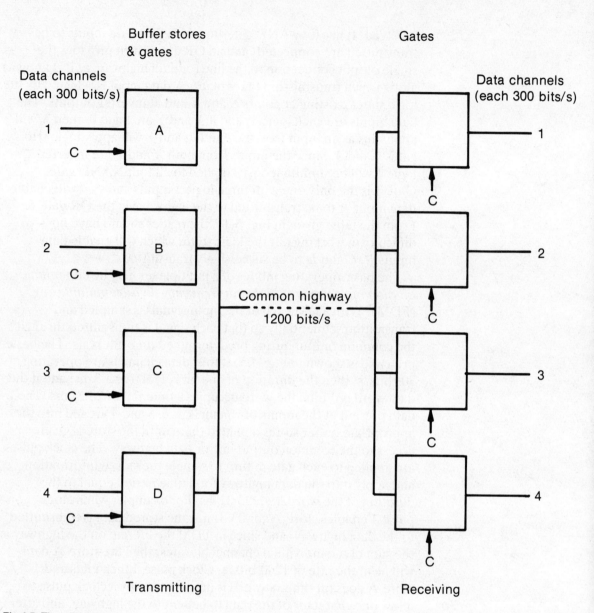

**Fig 4.5** Time division multiplexing. C = clock pulses

inputs to be connected to its single output, a multiplexer has several special applications such as parallel-to-serial data conversion, serial pattern generation, and the simplified implementation of Boolean functions.

**Demultiplexers** Basically these are the reverse of multiplexers, i.e., they have a single input and multiple outputs; the input can be connected to any

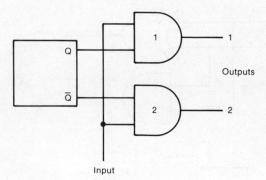

**Fig 4.6** Demultiplexer

one of the outputs. A simple 2-output circuit is shown in Fig. 4.6. The single input is applied to AND gates 1 and 2 and the flip–flop selects which gate is enabled. For example, when $Q$ is set, gate 1 only is enabled and so the input will be gated through to output 1, resetting the flip–flop. $\overline{Q}$ is then logic 1, gate 1 is inhibited, and so the input will now be gated through to output 2 via gate 2.

**Binary Adders**

These are combinational logic circuits used in digital computers, microprocessors and other digital equipment in which mathematical operations are to be performed. They are one of the many circuits in which the special characteristics of the exclusive OR gate are used. These characteristics are that its logical function is identical to the rules of binary addition, i.e., the inputs are the two single bits to be added, and its output is the single bit sum. The only function that this gate cannot handle is the carry function. An AND gate is therefore combined with it, as shown in Fig. 4.7(a), to produce a logic 1 carry output only when both inputs are logic 1. This circuit, however, is only a *half-adder* since it cannot add the carry input from an LSB position to the sum of inputs. To perform the function of adding multibit numbers, therefore, it must be combined with others of its kind to become a *full-adder* circuit, as in Fig. 4.7(b).

Gates 1 and 2 add the two inputs $A$ and $B$ and this sum is gated as an input to gate 3. The sum is added to the carry input from an LSB position, in the second half-adder circuit gates 3 and 4. The output of gate 4 is the correct sum. The carry outputs from gates 1 and 3 are ORed together in gate 5 to produce the correct carry output to the next MSB position.

**Operational Amplifiers**

Operational amplifiers ('op-amps') are very high-gain d.c. amplifiers used in circuits associated with analog measurements and functions, and in some cases with digital circuit applications. They

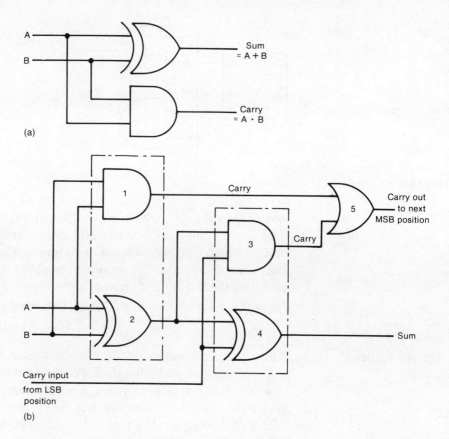

**Fig 4.7** Adders. (a) Half. (b) Full

are in integrated circuit form, and as Table 4.2 shows they can perform a number of functions according to whether their operating range is linear or non-linear.

The schematic symbols for an op-amp are shown in Fig. 4.8, and as an indication of the integrated circuitry which can be involved, an equivalent circuit of a dual op-amp for use as a summing amplifier or an integrator is also shown.

The negative input of an op-amp is called the *inverting* input, and the positive input is called the *non-inverting* input. If a positive input signal is applied to the inverting input with the non-inverting input grounded, the polarity of the output signal will be *opposite* to that of the input signal. If, on the other hand, a positive input signal is applied to the non-inverting input with the inverting input grounded, then the output signal polarity will be the *same* as that of the input signal. If the same signal is applied to both inputs, the two amplified output signals will be 180° out-of-phase and will completely cancel each other. Since in this case the amplifier responds only to differences between the two inputs, it is said to be a differential amplifier.

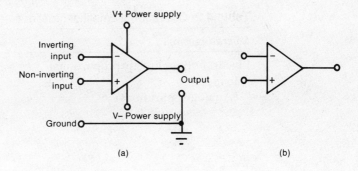
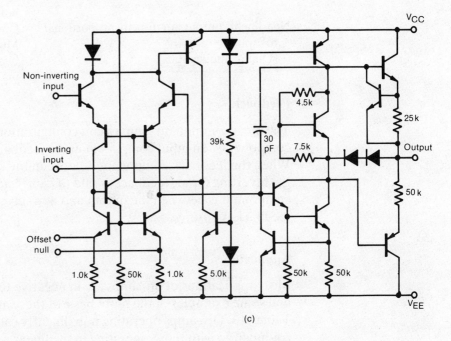
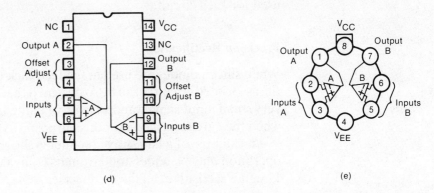

**Fig 4.8** Operational amplifier.
(a) Detailed schematic.
(b) Simplified schematic.
(c) Circuit. (d) DIL pack.
(e) Can pack.

Microelectronics in Aircraft Systems   75

**Table 4.2** Operational amplifier functions

| Operating range | Function |
|---|---|
| Linear: output directly proportional to input (Analog applications) | Inverting<br>Non-inverting<br>Integrating<br>Differentiator<br>Summer<br>Voltage follower<br>Precision Rectification<br>Logarithmic<br>Filters: high-pass, low-pass, band-pass, notch |
| Non-linear: output not directly proportional to input (Digital applications) | Comparator<br>Multivibrator: astable, bi-stable, monostable |

## Feedback

The most common op-amp circuit configuration uses two external components: an input component and a feedback component. When the feedback component is between the op-amp output and the inverting input, feedback is said to be *negative*, and when the component is between the output and non-inverting input, feedback is *positive*.

## Applications

Most applications of op-amps are in negative feedback circuits, and some examples of the more basic of these are summarized in Table 4.3. Op-amps operating non-linearly can be used in conjunction with those operating in the linear range; the input and output voltages are compatible, and the same power supply can be used for each circuit.

## Precision Rectification

When silicon diodes are used in the feedback loop of an op-amp, the associated circuitry can be so designed that it will precisely rectify very small input signals and so overcome the limitations of diodes when they are used alone for this purpose.

An example of a half-wave rectifier is shown in Fig. 4.9(a). Its operation may be understood from its equivalent circuit. When its $V_i$ signal is positive, all the feedback current $I_f$ will flow through $D_1$,

**Table 4.3** Op-amp applications

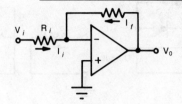

*Inverting:*
Output 180° out of phase with $V_i$. Input impedance just equal to $R_i$. Output impedance is zero since $V_o$ is determined by $I_f$, which is not affected by load.

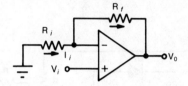

*Non-inverting:*
Output signal in phase with $V_i$. Since $V_0$ is referenced to grounded end of $R_i$, the voltage drop across $R_i$ is in series with output; $R_i$ and $R_f$ are therefore in series.

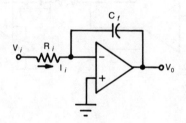

*Integrating:*
Takes the sum of $V_i$ signal over a period of time. Charging current of $C_f$ equal to current flowing in $R_i$. The value of $V_0$ is equal to linear voltage across $C_f$. The value of $V_0$ is proportional to negative of integral of input.

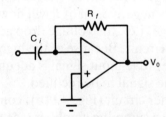

*Differentiating:*
Responds only to differences in $V_i$ which will result in current flowing through $C_i$ and $R_f$. The value of $V_0$ is equal to voltage drop across $R_f$ and proportional to negative of input derivative.

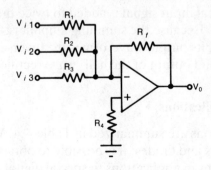

*Summing:*
Since negative feedback holds inverting input very close to ground potential, all input signals are electrically isolated from each other. $V_0$ is equal to inverted sum of input signals. $R_4$ is chosen as parallel combination of input and feedback resistances, assuming low source resistances.

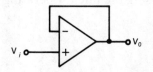

*Voltage-following:*
A non-inverting amplifier with $V_0$ fed directly back.

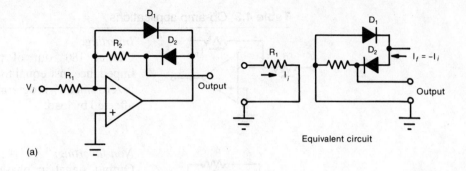

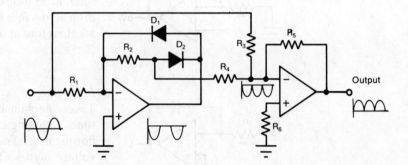

**Fig 4.9** Precision rectification.
(a) Half-wave. (b) Full-wave

and the output voltage of the circuit (not the op-amp) will be zero. When the $V_i$ signal is negative, all of $I_f$ will flow through $D_2$ and $R_2$, and an output voltage, i.e., the voltage drop across $R_2$, will appear at the output of the circuit. Because of the high gain of the op-amp, even a very small negative input signal is adequate to forward-bias $D_2$, thus allowing the signal to be rectified.

A full-wave rectifier circuit (Fig. 4.9(b)) consists of a half-wave rectifier followed by a summing amplifier (see also Table 4.3). Since the input resistors of the summing amplifier are selected with $R_3$ twice $R_4$, the original input signal is added to twice the output of the half-wave rectifier. Because the summing amplifier is also an inverting amplifier, its output waveform is the inverse of the sum of those at the input and output of the half-wave rectifier.

### Digital Circuit Applications

The basic applications are summarized in Table 4.4. With additional resistors, capacitors and diodes, it is possible to obtain a great number of variations in applications to special digital circuits, and indeed to circuits using the linear operating range of op-amps referred to earlier.

**Table 4.4** Op-amp applications (digital)

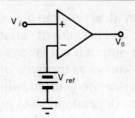

*Comparator:*
Compares $V_i$ to $V_{ref}$. When $V_i$ slightly greater than $V_{ref}$, op-amp swings into +ve saturation; $V_i$ slightly less than $V_{ref}$, it swings into −ve saturation. If $V_{ref}$ is ground, a comparator may be used to convert a sine wave to a square wave; thus sine wave +ve, op-amp swings into +ve saturation; sine wave −ve, op-amp swings into −ve saturation.

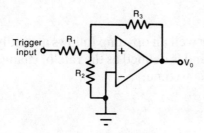

*Bistable multivibrator:*
In an op-amp multivibrator, the two stable states are −ve saturation and +ve saturation. Trigger input signal changes output from one state to the other. Op-amp thus operates as a flip-flop, the example shown being d.c. coupled, and with +ve feedback. When initially in +ve saturation, a +ve input pulse will have no effect. When −ve trigger pulse causes +ve input of op-amp to become −ve, $V_0$ also goes −ve, and the +ve feedback drives op-amp into −ve saturation. Changes to +ve saturation when +ve trigger pulse applied.

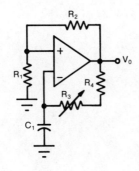

*Astable multivibrator:*
Generates a square-wave output as circuit switches between its two astable states. If op-amp in +ve saturation, the +ve input held at one-half saturation voltage by $R_1$ and $R_2$. Capacitor $C_1$ charges up through $R_3$ and $R_4$. When $C_1$ voltage exceeds half saturation voltage, the −ve input voltage becomes more +ve, and the op-amp swings into −ve saturation. The +ve input is then held at one-half −ve saturation voltage until $C_1$ charges to slightly less than one-half saturation voltage. Op-amp then swings back into +ve saturation and the cycle repeats.

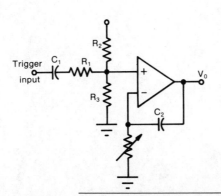

*Monostable multivibrator:*
Op-amp normally in +ve saturation due to −ve voltage at −ve input (through voltage divider $R_2$ and $R_3$). If a +ve trigger pulse is applied it causes $V_0$ to swing into −ve saturation. After swinging into −ve saturation, $C_2$ charges up through $R_4$, and as soon as +ve input becomes more +ve than −ve input, the op-amp automatically returns to +ve saturation.

### Offset

For a zero input, an ideal op-amp has exactly zero output. In practice, however, and because of variations in transistor characteristics, the output is not always zero. To compensate for this, a d.c. voltage known as the *input offset voltage* is applied to an amplifier. The compensated output, which is known as the *output offset voltage*, is also dependent on an *input bias current* which is the difference in d.c. bias currents of the inverting and non-inverting transistors used in an op-amp.

### Slewing Rate

This relates to how well an op-amp follows a rapidly changing input signal waveform, and is defined as the rate of change of rising voltage with respect to time.

# 5 Displays

The display of data is a requirement vital to the control of any aircraft and its associated systems, such display being traditionally presented by a variety of individual instruments in quantitative form, i.e., in terms of numerical values (Figs. 5.1 and 5.2). With the introduction, however, of microelectronics and digital signal processing technology into the avionic systems field, it was also a logical step to use the display techniques associated with this technology; in other words to present required data in alphanumeric form (both letters and numbers) and also pictorially or graphically (qualitative form). In fact the stage has already been reached whereby, for example, traditional scale and pointer type instruments designed to display data necessary for monitoring engine performance can be dispensed with entirely and replaced by a microprocessing method of 'painting' equivalent data on the screen of a cathode ray tube type of display unit.

Electronic, or to be more precise, optoelectronic† displays in avionic systems fall into two broad categories: those in which a single register of numeric values is required, and those requiring several lines of alphanumeric information or instructions, and/or display of situations in pictorial form. Typical examples of display technologies pertinent to these areas and their applications to avionic systems are given in Table 5.1.

The operation mode of displays may be either active or passive, the definitions of which are as follows.

*Active*

A display using phenomena potentially capable of producing light when the display elements are electrically activated.

*Passive*

A display which either transmits light from an auxiliary light source

†The production, utilisation and evaluation of electromagnetic radiation in the optical wavelength range and its conversion into electrical signals.

**Fig 5.1** Traditional flight instrument display

after modulation by the display device, or which produces a pattern viewed by reflected ambient light. Some passive displays can be either transmissive or reflective.

**Matrix Displays**

Displays of this type are usually limited to applications in which a single register of alphanumeric values is required and are based on what is termed a 7-segment matrix configuration, or in some cases,

**Fig 5.2** Traditional engine instrument display

**Table 5.1** Applications of electronic displays

| Display technology | | Operation mode | Typical applications |
|---|---|---|---|
| Matrix | light-emitting diode | Active | Digital counter displays of engine performance monitoring indicators; radio frequency selector indicators; distance measuring indicators; control display units of inertial navigation systems |
| | liquid crystal | Passive | |
| | gas discharge | Active | |
| | incandescent filament | | |
| Electron beam | Cathode ray tube | Active | Weather radar indicators; display of navigational data, engine performance, systems status, check lists |

a point or dot-matrix configuration. Examples of these are shown in Figs. 5.3 and 5.4 respectively. As will be noted from Table 5.1, the technologies appropriate to matrix displays are the light-emitting diode, the liquid crystal, gas discharge, and incandescent filament. We shall now deal with all but the last.

**Fig 5.3** Seven-segment display (courtesy Smiths Industries Ltd)

**Fig 5.4** Dot matrix display (courtesy Smiths Industries Ltd)

## Light-Emitting Diodes

Light-emitting diodes (LEDs) are essentially forward-biased p-n junction transistors formed from a slice or chip of gallium arsenide phosphide (GaAsP). When current flows through the chip it emits light which is directly in proportion to the current flow; light emission in the red, orange, green and yellow regions of the colour spectrum is obtained by proportioning of the constituent elements of the chip, and also by a technique of 'doping' with other elements, e.g., nitrogen.

In applying the LED technology to digital counter displays of avionic systems, both the 7-segment and dot-matrix configurations may be adopted. In the 7-segment configuration (Fig. 5.5) each of

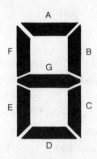

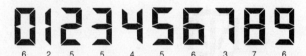

| No. of segments | 6 | 2 | 5 | 5 | 4 | 5 | 6 | 3 | 7 | 6 |

**Fig 5.5** Seven-segment display configuration

the segments is an individual chip mounted within a reflective cavity with a plastic overlay and a diffuser plate; the basic form of construction is shown in Fig. 5.6. The segments are formed as a sealed integrated circuit pack, the connecting pins of which are soldered to the associated printed circuit board. Depending on the application and the number of digits comprising the appropriate quantitative display, independent digit packs may be used, or combined in a multiple digit display unit.

The decimal numbers 0-9 are displayed by the illumination of two or more segments in a pattern that is generated by a combinational logic circuit known as a decoder, which converts binary coded decimal (BCD) numbers into the code required to operate the seven segments of the display. The letters which conventionally designate each of the segments, and the pattern generated by the decoder for the display of decimal numbers, is shown in Fig. 5.5.

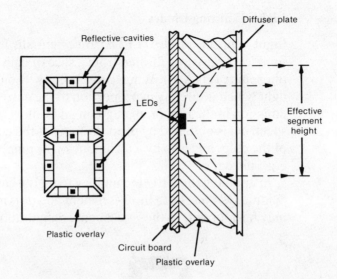

**Fig 5.6** Construction of seven-segment LED

A segmented configuration may also be used for the display of alphanumeric data, but this requires that the number of segments be increased from 7 to 13, and in some cases to 16, in order to display the full range of capital letters and numerals. Examples of this display, commonly referred to as 'starburst', are shown in Fig. 5.7, and from diagram (a) it should be noted how segments are illuminated to distinguish between the displays of the letter O and a

(a)  (b)

**Fig 5.7** Alphanumeric display
(a) 13-segment configuration
(b) 16-segment configuration

# HEXADECIMAL DISPLAY WITH LOGIC  TIL 311

SOLID STATE VISIBLE HEXADECIMAL DISPLAY WITH INTEGRAL TTL CIRCUIT TO ACCEPT, STORE, AND DISPLAY 4-BIT BINARY DATA

### description

This hexadecimal display contains a four-bit latch, decoder, driver, and 4 X 7 light emitting diode (LED) character with two externally driven decimal points in a 14 pin package. A description of the functions of the input of this device follows.

| FUNCTION | PIN NO. | DESCRIPTION |
|---|---|---|
| LATCH STROBE INPUT | 5 | When low, the data on the latches follow the data on the latch data inputs. When high, the data in the latches will not change. If the display is blanked and then restored while the enable input is high, the previous character will again be displayed. |
| BLANKING INPUT | 8 | When high, the display is blanked regardless of the levels of the other inputs. When low, a character is displayed as determined by the data in the latches. The blanking input may be pulsed for intensity modulation. |
| LATCH DATA INPUTS (A, B, C, D) | 3, 2, 13, 12 | Data on these inputs are entered into the latches when the enable input is low. The binary weights of these inputs are A - 1, B - 2, C - 4, D - 8 |
| DECIMAL POINT CATHODES | 4, 10 | These LEDs are not connected to the logic chip. If a decimal point is used, an external resistor or other current limiting mechanism must be connected in series with it. |
| LED SUPPLY | 1 | This connection permits the user to save on regulated $V_{CC}$ current by using a separate LED supply, or it may be externally connected to the logic supply ($V_{CC}$) |
| LOGIC SUPPLY ($V_{CC}$) | 14 | Separate $V_{CC}$ connection for the logic chip. |
| COMMON GROUND | 7 | This is the negative terminal for all logic and LED currents except for the decimal points. |

The LED driver outputs are designed to maintain a relatively constant on level current of approximately five milliamperes through each of the LED's forming the hexadecimal character. This current is virtually independent of the LED supply voltage within the recommended operating conditions. Driver current varies with changes in logic supply voltage resulting in a change in luminous intensity as shown in Figure 2. The decimal point anodes are connected to the LED supply, the cathodes are connected to external pins. Since there is no current limiting built into the decimal point circuits, this must be provided externally if the decimal points are used.

The TTL MSI chip is specially designed with a wider supply voltage range than standard Series 54/74 circuits so that it will operate from either a five volt or a six volt power supply.

The resultant displays for the values of the binary data in the latches are as shown below.

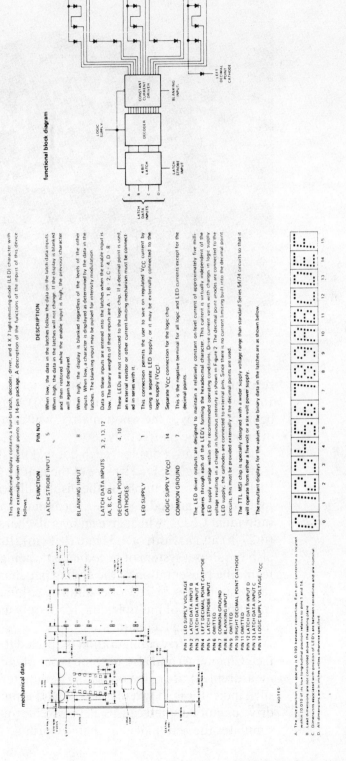

**Fig 5.8** Matrix LED display

Microelectronics in Aircraft Systems

**Fig 5.9** Percentage rpm indicator with a dot matrix display (courtesy Smiths Industries Ltd)

zero, the letter B and figure 8, and the letter D and zero. In the 16-segment configuration (diagram (b)) it is usual for the letters to be displayed over the whole matrix, while numerals are displayed over half of the matrix only.

In a dot-matrix display, each dot making up the decimal numbers is an individual LED. The size of the matrix governs the number of LEDs used; for example, in the display illustrated in Fig. 5.8 the matrix is designated $4 \times 7$, i.e., 4 columns and 7 rows, and uses 20 individual LEDs, which give an overall character height of 6·85 mm (0·27 in) and a width of 4·32 mm (0·17 in). Illumination of the LEDs also requires a system of decoding but, being arranged in matrix form, the presentation of the coding per character has to be done sequentially row by row or column by column. This is achieved by the application of multiplexing, read-only memory or microprocessing techniques.

The dot-matrix configuration is applied to CRT display systems (see Chapter 6) and also to some types of indicators required for the measurement of engine operating parameters. An example of one such indicator is shown in Fig. 5.9, and is unique in that the display drive circuit adopted causes an apparent 'rolling' of the digits which simulates the action of a mechanical drum-type counter as it responds to changes in measured parameters. The operating fundamentals of this instrument will be covered in Chapter 9.

**Liquid Crystal Display (LCD)**

This type of display, which is commonly used in digital watches and many types of pocket calculator, is also used in data displays for aircraft systems, e.g., for the display of selected radio frequencies (Fig. 5.10).

**Fig 5.10** Application of an LCD display

The basic structure of an LCD is shown in Fig. 5.11. It consists of two glass plates coated on their inner surfaces with a thin transparent conductor material such as indium oxide. The material on the front plate is etched into the standard display format of seven segments, each of which form an electrode. A mirror image of the digits with its associated electrical contact is also etched into the

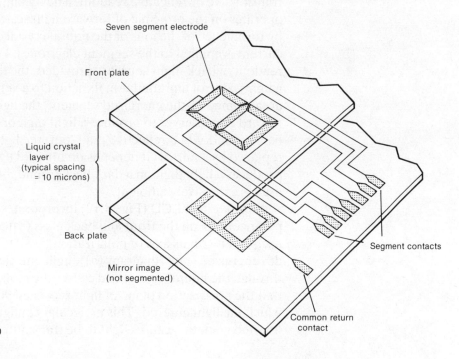

**Fig 5.11** Structure of an LCD

oxide layer of the back glass plate, but this is not segmented since it constitutes a common return for all segments. The space between the plates is filled with a liquid-crystal compound (esters and biphenyls are typical) which has a thread-like molecular structure, the molecules being oriented with their long axes parallel. The complete assembly is hermetically sealed with a special thermoplastic material to prevent contamination of the liquid-crystal compound by water vapour and oxygen.

When a low-voltage, low-current signal is applied to the segments from the decoder/drive circuits, the molecular order of the liquid-crystal compound is disturbed and this changes its optical appearance from transparent to reflective. The magnitude of the optical change, i.e., the contrast ratio, is basically a measure of the light reflected from, or transmitted through, the segment area, to the light reflected from the background area; a typical ratio is 15:1. Thus, an LCD does not emit light, but merely acts on light passing through it. Energizing of the segments is accomplished by the simultaneous application of a symmetrical out-of-phase signal to the front and back electrodes of a segment, which thereby produces a net voltage difference. When two in-phase signals are applied to the display segments, the net voltage is zero and the display segments spontaneously relax to the de-energized state.

An LCD may be of either the dynamic-scattering type or the field-effect type, and these in turn may produce either a transmissive or reflective readout. The dynamic-scattering display operates on the principle of forward light scattering, which is caused by turbulence of the ions of the liquid-crystal compound when current is applied to the segment electrodes. For a transmissive readout, a back-light source is provided, the light being directed by a light-control film similar in its action to a venetian blind. In the area defined by the energized segment, the light is then scattered up toward the observer to produce a light digit or character on a dark background. For a reflective read-out, the light-control film is replaced by a mirror; it depends on forward light scattering and also produces a light digit on a dark background, but the light source used is extensively ambient.

A field-effect LCD (Fig. 5.12) incorporates additional layers of polarizer film on the front and back glass plates of the assembly, each layer with its axis of polarization at 90° to the other. In the de-energized state (diagram (a)), light entering the display passes through the liquid-crystal and causes the molecules to realign such that the polarization plane of light is twisted 90° to the plane in which the light entered. This molecular configuration, also known as 'twisted nematic', causes light to be transmitted through the

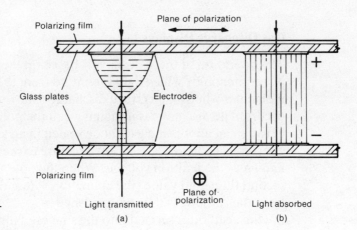

**Fig 5.12** Operation of a field-effect LCD

complete LCD, and the display would appear as a light area. When the LCD is energized (diagram (b)), the liquid crystals 'untwist' so that incident light passes through in the same plane of polarization. The rear polarizer film, however, will now absorb the light so that the display will appear as a dark area.

For a transmissive field-effect display the polarizers are oriented so that the front and back films are parallel, and this results in the appearance of a light digit or character on a dark background, as shown in Fig. 5.13(a). In the reflective configuration, the polarizers are orthogonal to each other and in addition a diffuser reflector is attached to the back of the liquid-crystal cell structure. This arrangement produces a dark digit or character on a light background, as in Fig. 5.13(b).

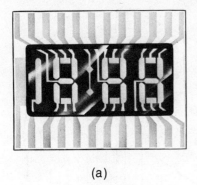

(a)

(b)

**Fig 5.13** Field-effect LCDs

Field-effect LCDs are more widely used than displays based on dynamic-scattering by virtue of such factors as low-power consumption, better display contrast, reliability, and longer life.

## Gas Discharge Displays

The operation of these displays is based on the well-known phenomenon by which light is emitted from the excitation of gas molecules when an electrical discharge (known as glow discharge) occurs. The discharge is initiated when the potential difference between an anode and a cathode sealed in an atmosphere of an inert gas (typically neon) at very low pressure exceeds a critical value, known as the ignition voltage. When the potential drops below a second threshold value (the extinction voltage) the glow discharge from the cathode is then extinguished.

Thus, cathodes formed into the shapes of alphanumeric characters can display the de-coded data. This is a display technique often applied to some types of workshop test equipment, where the electrodes are contained in a sealed glass envelope resembling a thermionic tube. In the application of gas-discharge displays to airborne equipment, however, the cathodes are in the 7-segment configuration and operate against a common anode.

A further development of this type of display is the one known as the plasma panel, the basic construction of which is shown in Fig. 5.14. The panel is divided into a display section consisting of a glass panel and transparent display anodes, and a scanning section

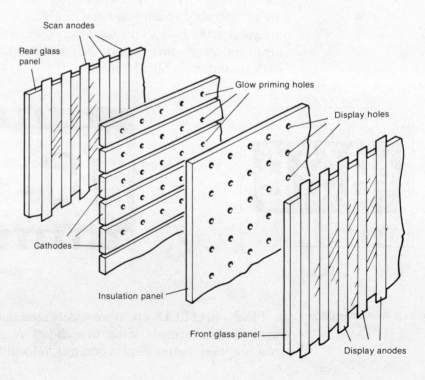

**Fig 5.14** Plasma panel display

consisting of a rear glass panel and non-transparent scan anodes. Sandwiched between the plates is an insulation panel containing holes which define gas discharge cells, and cathodes corresponding in number to that of the anodes. The cathodes are positioned at right angles to both sets of anodes, and like the insulation sheet they contain holes in which priming of the glow discharge takes place. The whole assembly is sealed to contain a gas mixture which is predominantly neon. In operation, a discharge is made to travel down troughs between the cathodes and the scan anodes by a clock generator which also drives the cathodes. The light generated by this process is not visible from the front display panel. The priming holes in the cathodes allow excited atoms to penetrate the display holes in the insulation panel from the rear-scan glow when the display anodes are energized. The glow in the display holes can be seen from the front glass display panel. By selectively energizing the display anodes and using conventional time-division multiplexing techniques, it is possible to cause any of the display holes to glow. Normally, the display anodes are driven by information stored in a read-only memory so that alphanumeric characters are displayed in dot-matrix format.

## Encoding and Decoding

Display devices are, of course, designed for use with digital systems, and to operate them the data relevant to the functions of the systems must first be arranged in a coded format (encoded). These data must then be detected and translated into an intelligible display format (decoded). It is necessary, therefore, to incorporate elements or sections which will perform these functions within display control circuits.

The design of the encoding section depends on the system in which the display is used; for example, it depends on whether the display is to be simply of decimal numbers encoded via manual selection of correspondingly numbered keys on a keyboard, or of numeric data associated with such parameters as engine speed, temperatures, distance flown, etc., which are subject to changes automatically encoded via sensors or monitoring circuits. A decoding section, on the other hand, particularly where the 7-segment display format is concerned, is of a more standard form.

As a basis for understanding how a display device operates in response to encoded and decoded information, it is useful to take as an example that section of a digital system familiar to almost everybody nowadays, namely, the pocket calculator. The principal sub-systems which make up the display section are shown in

Fig. 5.15. With the exception of the display and keyboard, all the sub-systems are contained within the chip of an integrated circuit pack.

Figure 5.16 is a schematic development of Fig. 5.15, and serves not only to show how the segments of a display are illuminated, but also as a further exercise in the understanding of logic gate

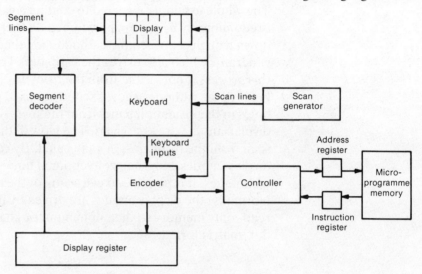

**Fig 5.15** Encoding and decoding sub-systems

applications. Let us assume that the number 5 is to be displayed. With the calculator switched on, the scan generator within the IC pack supplies power to the nine scan lines connected to the correspondingly numbered keys, one at a time and thousands of times per second. From Fig. 5.16(a) it will be noted that when the number 5 key is pressed, its associated switch connects scan line 5 to a keyboard line, allowing current to flow through it to the keyboard encoder. The remaining scan lines are similarly connected to the encoder when their corresponding keys are pressed.

The keyboard encoder is in two decision-making sections: the first section, of nine NAND gates, decides which number key has been pressed, according to which scan line is 'on'. The encoded answer is transmitted by turning 'on' one of ten number lines (number 0 not shown on the diagram makes up the tenth) leading to the second section; this is of four NAND gates, and decides which of the four wires leading to the display register to turn on in order to transmit a keyed number according to the 8421 binary number code. Thus, in our example, scan line 5 is 'on', and transmits a logic 1 pulse (positive logic) to NAND gate 5 together with the logic 1 pulse in the keyboard input line. The output from NAND gate 5 is, therefore, at logic 0, and since during its scanning sequence the scan generator finds the other scan lines open, the number 5 is first

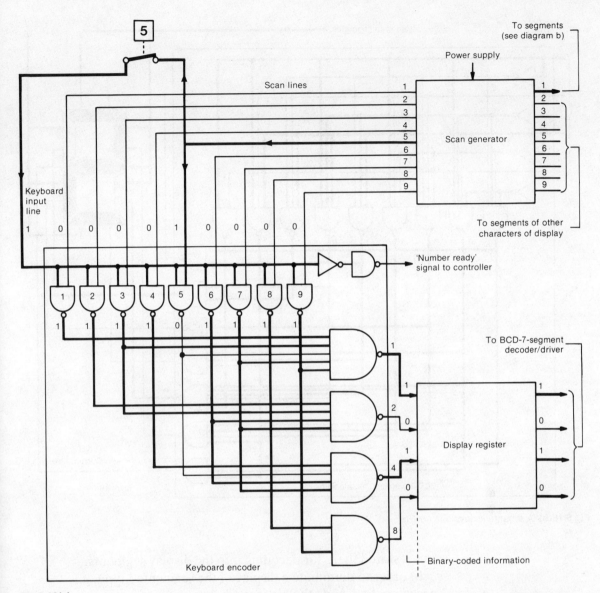

**Fig 5.16(a)**

encoded as 1000010000, and then converted by the NAND gates to 111101111 (Fig. 5.16(a)). This code is then transmitted to the four NAND gates in the second section of the encoder. The output from the four gates is further encoded in accordance with the 8421 binary code, and as will be recalled, in this code the corresponding decimal numbers are the sum of the numbers represented by lines which are 'on' and transmitting a logic 1. As shown in Fig. 5.16(a), line 4 plus line 1 are both transmitting logic 1, and so the keyed decimal number 5 is supplied to the display register in the binary code 0101,

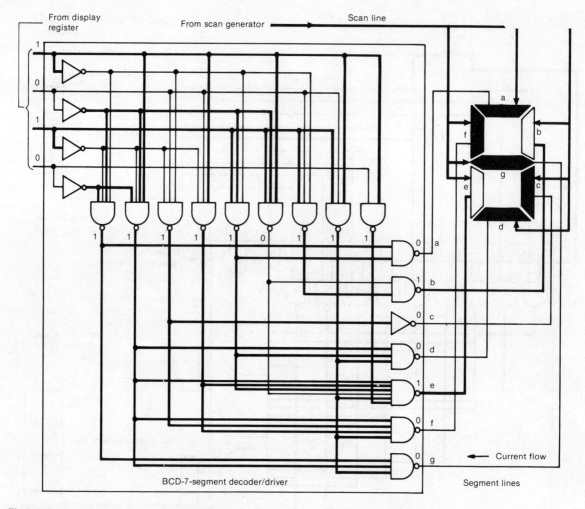

**Fig 5.16(b)** A schematic development Fig 5.15

and is stored there for decoding into the display segments.

In order to illuminate a display of the segmented type, it is necessary to decode the binary coded decimal information stored in the display register; in other words to convert from a 4-bit code to a 7-bit code. This is performed by a device known as a BCD-7-segment decoder/driver, the gating arrangement of which is shown in Fig. 5.16(b). The input side of the decoder is coupled to the display register and, as will be noted, there are nine NAND gates which are coupled to eight input lines, four of which invert the display register output. The NAND gates decode these inputs as 111110111 and transmit this code to the segment driver section made up of NAND gates and a buffer, and they provide the 7-bit code 0100100 which, as indicated in the decoder truth table of

96 Displays

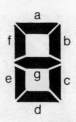

| Input combinations | | | | Output combinations to segments | | | | | | | |
|---|---|---|---|---|---|---|---|---|---|---|---|
| D 8 | C 4 | B 2 | A 1 | a | b | c | d | e | f | g | |
| 0 | 0 | 0 | 0 | 0 | 0 | 0 | 0 | 0 | 0 | 1 | 0 |
| 0 | 0 | 0 | 1 | 1 | 0 | 0 | 1 | 1 | 1 | 1 | 1 |
| 0 | 0 | 1 | 0 | 0 | 0 | 1 | 0 | 0 | 1 | 0 | 2 |
| 0 | 0 | 1 | 1 | 0 | 0 | 0 | 0 | 1 | 1 | 0 | 3 |
| 0 | 1 | 0 | 0 | 1 | 0 | 0 | 1 | 1 | 0 | 0 | 4 |
| 0 | 1 | 0 | 1 | 0 | 1 | 0 | 0 | 1 | 0 | 0 | 5 |
| 0 | 1 | 1 | 0 | 0 | 1 | 0 | 0 | 0 | 0 | 0 | 6 |
| 0 | 1 | 1 | 1 | 0 | 0 | 0 | 1 | 1 | 1 | 1 | 7 |
| 1 | 0 | 0 | 0 | 0 | 0 | 0 | 0 | 0 | 0 | 0 | 8 |
| 1 | 0 | 0 | 1 | 0 | 0 | 0 | 0 | 1 | 0 | 0 | 9 |

**Fig 5.17** Truth table for a BCD-7-segment decoder

Fig. 5.17, corresponds to decimal number 5. The driver section is connected to each segment of the display through appropriate segment lines. The scan line 5 is also connected to each of the display segments, and as the scan generator outputs are positive logic, current will be supplied to each of the segments because scan line 5 is at logic 1. The driver output from the BCD-7-segment decoder is, however, using negative logic, and so current will flow through the display to illuminate segments 'a', 'c', 'd', 'f' and 'g', which for the display of decimal number 5 are the only ones connected to segment lines at logic 0.

To check that all segments of a display are functioning correctly, it is necessary to select the decimal number 8 key. With the aid of Figs. 5.16 and 5.17 the reader may find this a useful exercise to trace out.

**Counting**

Many of the conventional instruments in current use feature a display format similar to those shown in Fig. 5.18, i.e., a single

**Fig 5.18** Conventional counter display of instruments (courtesy Smiths Instruments Ltd)

pointer which is driven via a servomechanism, in response to signals from a transducer, and a drum type decade counter which is mechanically coupled to the servomechanism. In operation, the drums rotate each other through specific drive ratios to count up and down throughout the appropriate measuring range. This mechanical form of digital counting also has its equivalent in microelectronics and takes the form of counters which are designed to generate the necessary codes for driving matrix displays.

Two of the most widely used counters are the *binary* which counts in pure binary code, and the *BCD* which counts in the standard 8421 BCD code referred to earlier. In each case the circuit is made up of flip–flops connected in cascade, the number of flip–flops used being determined by the maximum count capability required. Figure 5.19 shows the connection and waveforms of a 4-bit binary counter using four J–K flip–flops and having a maximum count capability of $N = 15$ (binary 1111). The formula used for determining this is $N = 2^n - 1$, where $n$ is the number of flip–flops.

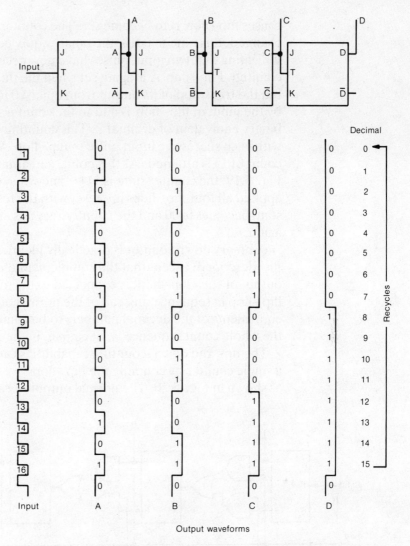

**Fig 5.19** 4-bit binary counter and waveforms

The input pulses to be counted are applied to the toggle (T) input of flip–flop A; while the inputs to the other three flip–flops are from the normal outputs of the preceding flip–flop. The JK inputs to each are 'open' or logic 1. If the counter is initially reset, the normal outputs of the four flip–flops will be binary 0. When the first input pulse occurs, flip–flop A will be set, and since a flip–flop changes state each time a trailing-edge transition occurs on its input waveform, the counter state will therefore be changed. Since flip–flop A corresponds to the least significant bit of the word, then in reading from right to left in Fig. 5.19 we note that the 4-bit word at the count of the first pulse is 0001, which is the binary equivalent of decimal 1. When the second input pulse occurs flip–flop A is now reset, and its normal output switches from binary 1 to binary 0 and causes flip–flop B to become set. The counter state is again changed, this time to 0010, the binary equivalent of decimal 2, indicating that two input pulses have now occurred and have been counted. Flip–flop A is again set when the third input pulse occurs, but the transition of its output from binary 0 to binary 1 is ignored by the input of flip–flop B and so the counter state is now 0011, the binary equivalent of decimal 3. This counting-up process continues with each successive input pulse to flip–flop A until the maximum count of 15 is attained. At this point, and as may be seen from Fig. 5.19, the counter state is 1111, and when the next pulse is applied all four flip–flops are reset with the result that the counter state becomes 0000 and the counter recycles and starts to count again.

A binary down counter is practically identical to the up counter just described, except that the complement output, i.e., the negated output of each flip–flop, is connected to the toggle input of the next flip–flop in sequence, instead of the normal output. The binary equivalents of the decimal numbers to be counted are the same, but the whole count sequence is, of course, in the reverse order.

The up- and down-counting capabilities can be combined within a single counter by coupling the flip–flops via AND and OR gates, as shown in Fig. 5.20. The normal output of each flip–flop is applied

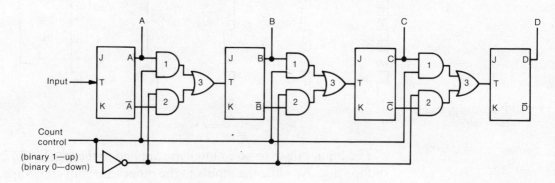

**Fig 5.20** Binary up/down counter

to each gate 1, while the complement output of each flip–flop is applied to each gate 2; the gates therefore determine whether the normal or the complement signals toggle the next flip–flop in the sequence. The other inputs to the gates are supplied from count control lines which determine whether the counter counts up or down. If the count control input is binary 1, i.e., the counter is counting up, the gate 1s are enabled. The normal output of each flip–flop is then coupled through gates 1 and 3 to the input of the next flip–flop in sequence. During the counting-up process, all gate 2s are inhibited. When the count control input is binary 0, the gate 2s are enabled, thereby coupling the complement output of each flip–flop to the next one in sequence, through gates 2 and 3. With this arrangement the counter counts down.

A BCD counter operates in a similar manner to a pure binary counter, but since it is designed to have only ten discrete states representing the decimal numbers 0 to 9, differences in the interconnections between flip–flops are necessary. As will be noted from Fig. 5.21, these differences consist principally of a feedback

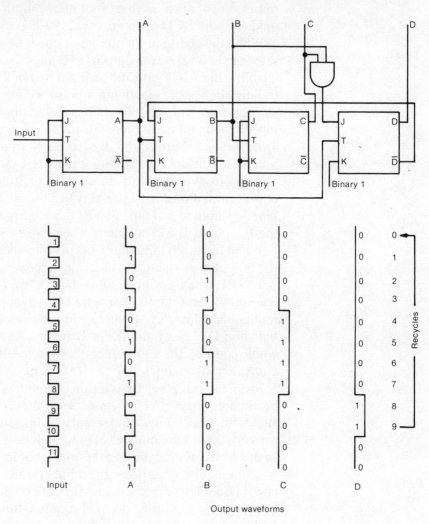

**Fig 5.21** BCD counter and waveforms

path from the complement output of flip–flop D to the J input of flip–flop B, and the inclusion of an AND gate to monitor the output states of flip–flops B and C, and to generate a control signal for operating the J input to flip–flop D. If we assume that the counter is initially reset, then the binary outputs of flip–flops B and C will be 0, and as they are supplied as inputs to the AND gate, the output of the gate and the J input of flip–flop D will also be 0. The complement output of flip–flop D, which is binary 1 during the reset state, will therefore enable the flip–flop B and permit it to toggle when flip–flop A changes state. If count pulses are now applied, the states of the flip–flops will start changing in the same sequence as a normal binary counter. It will, however, be noted from Fig. 5.20 that this is only the case for the first eight input pulses; the operations that occur on the ninth and tenth pulses are unique to the BCD counter.

When the eighth input pulse is applied, flip–flops A, B and C are set while D is reset, the outputs of B and C are binary 1 thereby enabling the AND gate and an input to J of flip–flop D. When the trailing edge of the eighth input pulse occurs, the counter state changes from 0111 to 1000. In this new state, the outputs from flip–flops B and C are binary 0, and so the J input to flip–flop D is again 0. Since there is a binary 1 at the K input of D, this flip–flop is set and conditions are right for it to be reset when an input at T switches from binary 1 to binary 0. In addition, the complement output from flip–flop D is binary 0 at this time, thereby keeping a binary 0 input at J of flip–flop B; the occurrence of a clock pulse at the T input of B will not affect it because it too is reset.

When the ninth input pulse occurs, flip–flop A sets, and as this is the only change that takes place the binary number in the counter is now 1001. The transition of flip–flop A from binary 0 to binary 1 is ignored by the T input of flip–flop D. On application of the tenth input pulse, flip–flop A will toggle and reset and it will also cause flip–flop D to reset; flip–flops B and C also remain reset. The whole counter, therefore, recycles from the 1001 state (the binary equivalent of decimal 9) to the 0000 state.

From the foregoing description of counter operation, it will be apparent that by connecting the counter outputs to a BCD-to-7-segment decoder and appropriate display, an automatic presentation of the number of input pulses applied can be obtained. In order to cater for the greater range required in more practical measurement applications, the number of BCD counters is correspondingly increased, and they are connected in cascade. The basic arrangement for a decimal counting function by means of an LED display is shown in Fig. 5.22; the maximum count in this case would be 999.

The outputs at A and D of each counter are respectively the *least significant bit* and the *most significant bit* of the words contained in them. Similarly, the equivalent digits in decimal will be the least significant digit in the input counter (number 1) and the most

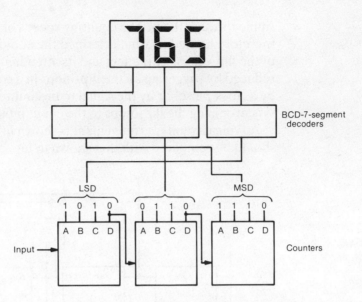

**Fig 5.22** Cascading of BCD counters

significant digit in counter number 3. As input pulses are applied to counter number 1, the counter will be incremented in the manner already described. When ten pulses have been counted, and the counter recycles, counter number 2 will now be triggered by an input from D of counter number 1, and will therefore count the binary equivalent of decimal 1. After one hundred input pulses to counter number 1, ten will have been supplied to counter number 2, and so it will have recycled to trigger counter number 3 and cause it to count the binary equivalent of decimal 1. It is clear that counter number 3 will recycle after one thousand input pulses at counter number 1.

**Frequency Dividing**

A binary counter is also a frequency divider, and this may be seen from the waveforms in Fig. 5.19. If, for example, the input to a 4-bit counter is a 100 kHz square wave, the outputs of flip–flops A, B, C and D would be 50, 25, 12·5 and 6·25 kHz respectively; i.e., the outputs of each flip–flop are one-half of the input frequency. Since the output of a counter is always some sub-multiple of two, for a 4-bit counter this value would be 16; when this is divided into the input frequency, the counter output frequency would therefore be 6·25 kHz.

**Synchronous Counters**

The binary counters described earlier are known as asynchronous or ripple counters, because the flip–flops are not controlled by a single common pulse and, being connected in cascade, the count pulses ripple through them. The counting speeds of such counters are therefore limited by what is termed the additive propagation delay of the flip–flops. This effect and its attendant errors in counting are reduced by triggering all the flip–flops in a counter simultaneously by a clock pulse, or by the signal to be counted; in other words, synchronizing the flip–flops to the count input.

A typical counter arrangement is shown in Fig. 5.23, and this should be compared with that shown in Fig. 5.20. The J–K inputs

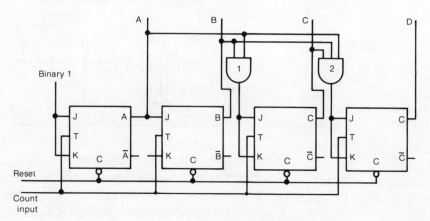

**Fig 5.23** Synchronous binary counter

on flip–flop A are connected to binary 1 so that it is permanently enabled, and it toggles or changes state in the same way as that of a ripple counter. Since the J–K inputs to flip–flop B are controlled by the normal output of A, then B changes state only when the output of A is binary 1. It will be noted that the output signals from flip–flops A and B are supplied to an AND gate (1) and its output is used to control the J–K inputs on flip–flop C which will change state at the first count pulse occurring after A and B are binary 1. The output signals from A and B, and from flip–flop C, are supplied to a second AND gate whose output then controls the J–K inputs on flip–flop D. The counting sequence and waveforms of this arrangement are identical to those given in Fig. 5.20, and apart from the mode of operation just described, it is still a binary counter.

# 6 Cathode ray tube displays

Displays based on the electron beam scanning technique as adopted in cathode ray tubes have been in use in aircraft for very many years. For example, during World War II military aircraft used equipment developed from ground-based radar systems. Depending on an aircraft's specific operational role, crews were able to navigate by 'radar mapping' of terrain, to identify ground target areas, and to detect the positions of hostile intercepting aircraft. Development of such equipment has, of course, continued with the development of military aircraft themselves to fulfil ever-changing strategic requirements. It can, for obvious reasons, only be stated that some highly sophisticated electronic display equipment is now in use.

As far as civil aircraft are concerned, the application of electronic display technology has until recent years been limited to the indicators of weather radar systems. This situation has, however, also undergone considerable change largely as a result of systems analysis, exploration of the versatility of the cathode ray tube (CRT), and investigation into methods whereby all associated data could be programmed into microprocessors and computers. These reached such high levels of sophistication and capacity for data processing that it became possible for a single CRT indicator unit, under microprocessed computer control, to project the same quantity of information which otherwise has to be provided by a large number of individual conventional type instruments. Another factor having considerable influence on the development of CRTs (or electronic displays as they are commonly called) for large civil transport aircraft operations resulted from investigations into ways in which flight deck layouts could be improved so as to reduce flight-crew workloads while still enhancing flight safety.

Electronic display systems are now extremely wide-ranging in respect of the information they can provide. The first three civil transport aircraft to adopt such display systems were the Boeing 757, 767 and Airbus A310. The flight-deck layouts of two of these aircraft are shown in Figs 6.1 and 6.2.

**Fig 6.1** Flight deck layout of the Boeing 757 (courtesy British Airways)

**Principle of the CRT**

A CRT (Fig. 6.3) is a thermionic device which consists of an evacuated glass envelope, inside which are positioned an electron gun and beam-focusing and -deflection systems. The inside surface of the screen is coated with a crystalline solid material known as a phosphor. The electron gun consists of an indirectly heated cathode biased negatively with respect to the screen, a cylindrical grid surrounding the cathode, and two (sometimes three) anodes. The grid is maintained at a negative potential, its purpose being to control the current and so modulate the beam of electrons passing

**Fig 6.2** Flight deck layout of the Airbus A310 (courtesy Airbus Industrie)

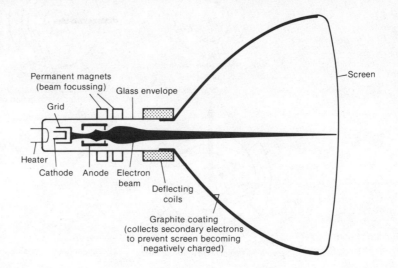

**Fig 6.3** Cathode ray tube

Microelectronics in Aircraft Systems 107

through the hole in the grid. The anodes are at a positive potential with respect to the cathode, and they accelerate the electrons to a high velocity until they strike the screen. The anodes also provide a means of focusing and, as will be noted from Fig. 6.3, this is in two stages.

The forces exerted by the field set up between the grid and the first anode bring the electrons into focus at a point just in front of the anode, at which point they diverge, and are then brought to a second focal point by the fields in the region between the three anodes. A focus control is provided which by adjustment of the potential at the third anode makes the focal point coincide with the position of the screen. When the electrons impact on the screen, the phosphor material luminesces at the beam focal point causing emission of a spot of light on the face of the screen.

To 'trace out' a luminescent display it is necessary for the spot of light to be deflected about horizontal and vertical axes, and for this purpose a beam-deflection system is also provided. Deflection systems can be either electrostatic or electromagnetic, the former being used in oscilloscope type instruments, and the latter in tubes used in television receivers, video display units, and other equipment requiring the display of alphanumeric data.

The manner in which an electromagnetic field is able to deflect an electron beam is illustrated in Fig. 6.4. A moving electron constitutes an electric current, and so a magnetic field will exist around it in the same way as a field around a current-carrying

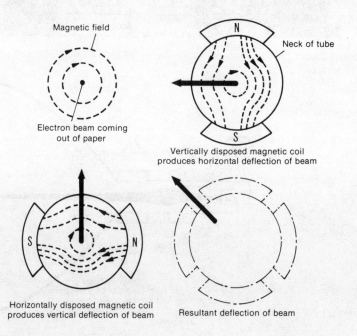

**Fig 6.4** Electron beam deflection

conductor. In the same way also that a conductor will experience a deflecting force when placed in a permanent magnetic field, so a CRT electron beam can be forced to move when subjected to electromagnetic fields acting across the space within the tube. Coils are therefore provided around the neck of the tube, and are configured so that fields are produced horizontally ($X$-axis fields) and vertically ($Y$-axis fields). The coils are connected to the signal sources whose variables are to be displayed, and the electron beam can be deflected to the left or right, up or down, or along some resultant direction depending on the polarities produced by the coils, and on whether either one alone is energized, or both are energized simultaneously.

## Colour CRT Displays

Colour CRT displays are now widely used in weather radar indicators, and are the norm for the CRT indicators designed for the display of navigational data, engine operating parameters, and other flight management data associated with the systems developed for such aircraft as the Boeing 757, 767 and Airbus A310. In these latter systems the display of weather data is also integrated with the other data displays, and since there is a fundamental similarity between the methods through which they are implemented, the operation of a weather radar indicator serves as a useful basis for study of the display principles involved.

The video data received from a radar antenna is conventionally in what is termed rho–theta form corresponding to the sweep of the antenna as it is driven by its motor (Fig. 6.5(a)). In the earlier types

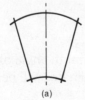
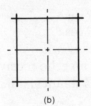

**Fig 6.5** Data cells. (a) Rho–theta. (b) X–Y

(a)  (b)

of monochrome indicators the display itself also corresponds to this rho–theta sweep, and occurred for each group of pulse repetition frequency pulses as the antenna scanned through the sector. In a colour display indicator, however, scanning is somewhat similar to that adopted in the tube of a television receiver or video display monitor, i.e., *raster* scanning in horizontal lines. The received video data is still in rho–theta form, but in order for it to be displayed it must be converted into an $X$–$Y$ co-ordinate format, as shown in Fig. 6.5(b). This format permits the display of other data in areas of the screen where the weather is not being displayed. In addition it permits a doubling up of the number of data cells as indicated by the dotted lines.

Each time the transmitter transmits a pulse, the receiver begins receiving return echos from targets at varying distances from the transmitter. This data is digitized to provide output levels in binary form and is supplied to the indicator on two data lines. The two bits of binary data can represent four conditions corresponding to the level of the return echos, which in turn are related to the weather prevailing at the range in nautical miles selected. The binary data and the four conditions are as follows:

00   Zero or low-level returns below the detectability of the radar system
01   Low returns from areas of lowest rainfall rate
10   Moderate returns from areas of moderate rainfall rate
11   Strong returns from high-density rainfall areas

The data are stored in memories which, on being addressed as the CRT is scanned, will at the proper time permit a display of the conditions. The four conditions are displayed as a blank screen, and green, yellow and red areas respectively.

**Scan Conversion**

As noted earlier, conversion of the data in rho–theta form must be converted to an $X-Y$ coordinate scan, and the principle of this is shown in Fig. 6.6. With a target at point P, at a range $R$ and an angle $\theta$ it will have coordinates:

$X = R \sin \theta$
$Y = R \cos \theta$

Thus an echo received at an azimuth angle of, say, 30° and at a range of 235 nautical miles, the coordinates will be:

$X = 235 \sin 30° = 117 \cdot 5$ nautical miles
$Y = 235 \cos 30° = 203 \cdot 5$ nautical miles

The conversion is performed by a microprocessor on the display circuit board within the indicator.

**Screen Format**

The coordinate system format of the screen is shown in Fig. 6.7, and from this example it will be noted that the screen is divided into two halves representing two quadrants in the coordinate system. The

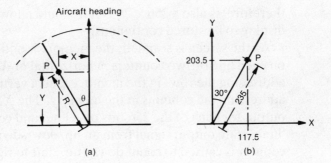

**Fig 6.6** Scan conversion

origin is at the bottom centre, so that values of $X$ are negative to the left and positive to the right; all values of $Y$ are positive. The screen is scanned in 256 horizontal lines, and there are 256 bits of information displayed on each line. Each line is located by a value of $Y$ and each bit by a value of $X$; the screen therefore has a $256 \times 256$ matrix. The $X$ and $Y$ values are used to address the memory and display the information stored there as the appropriate time in the scan occurs. The memory for the weather data is in two parts which store the MSBs and LSBs of the data words that represent the colours red, yellow or green and the corresponding weather conditions referred to earlier. Each part of the memory contains one address for every bit on every line in the display; each memory,

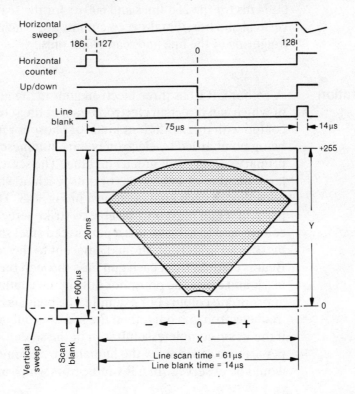

**Fig 6.7** Screen format

Microelectronics in Aircraft Systems   111

therefore, is also a 256 × 256 matrix, and allows the entire weather display to be stored continuously.

As the screen is scanned, the memory is addressed at each point on each line by two counters; a horizontal or $X$ counter for addressing the rows in the memory, and a vertical or $Y$ counter for addressing the columns in the memory. The $X$ counter generates an output for each of the 256 bits on a line, and counting is started by a 'high' state output signal from an up/down divider circuit. The counter is caused to count down (i.e., left to right) from the number 186 to 0 at the centre of the screen. When it reaches 0, the divider circuit changes to a 'low' state output, thereby causing the counter to count up to the number 128 at the end of the line, at which point a 'line blank' pulse of 14 μs duration is generated. The line scan time is about 61 μs, and so the total time for each line is 75 μs. The divider circuit again changes to a 'high' state to cause the counter to start counting down for the next line. This process is repeated for all the remaining lines.

An output from the $X$ counter is also applied to the $Y$ counter, which counts to 256 (one for each of the lines) plus eight counts for a scan blank time to allow for the CRT spot to return to the upper left corner of the screen. This process is repeated, and since there are 256 lines in the display it takes 20 ms to scan the entire screen (19·4 ms for the 256 lines and 600 μs for the scan blank time). The vertical and horizontal sweep circuits are synchronized by the triggering of the line and scan blank pulses.

**Colour Generation**

A colour CRT has three electron guns in the neck of the tube, each of which can direct an electron beam at the screen. The screen is coated with three different kinds of phosphor material which, on being bombarded by electron beams, luminesce in each of the three primary colours red, green and blue. The screen is divided into a large number of small areas or dots, each of which contains a phosphor of each kind, as shown in Fig. 6.8. The beam from a particular gun must only be able to strike screen elements of one colour, and to achieve this a perforated steel sheet called a *shadow mask* is accurately positioned adjacent to the coating of the screen. Beams emitted from each gun pass through the perforations in the mask and cause the phosphor dots in the coating to luminesce in the appropriate colour. For example, if a beam is being emitted by the 'red' electron gun only, then the red dots only would luminesce, and if the beam completes a full scan of the screen, then as a result of persistence of vision by the human eye a completely red screen would be observed. In CRT indicators which also display

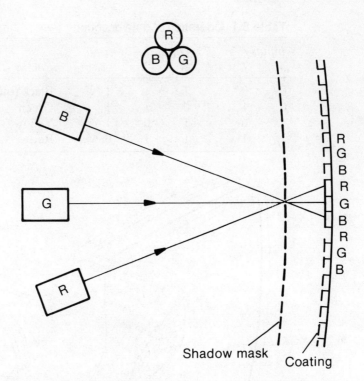

**Fig 6.8** Shadow mask tube principle

alphanumeric data other colours are required, and these are derived by independent control of the three guns and of their beam currents, so that as the beams strike the corresponding phosphor dots the basic process of mixing of primary colours takes place. For example, if all three guns are emitting, a mix of red, green and blue phosphor luminescence will produce a white scan or trace, and similarly if the red and blue guns are emitting the mix of red and blue phosphors will produce a mauve scan or trace.

A section view of a colour CRT is shown in Fig. 6.9.

Referring once again to the weather radar indicator application, we can see that the data read out from the memory, apart from being presented at the appropriate location of the CRT screen, must also be displayed in the colours corresponding to the weather prevailing. In order to achieve this, the data is decoded to produce outputs which, after amplification, will turn on the requisite colour guns; the data flow is given in Fig. 6.10. The memory output is applied to a data demultiplexer whose output corresponds to the MSBs and LSBs (M and L) of the 2-bit binary words and is supplied to a data decoder. The inputs are decoded to provide 3-bit output words corresponding to the colours to be displayed, as shown in Table 6.1. The outputs are then applied to the colour decoder and priority encoder circuit, and this in turn provides three outputs each

**Table 6.1** Operation of data decoder

| Inputs | | Outputs | | | |
|---|---|---|---|---|---|
| M | L | $A_1$ | $A_2$ | $A_3$ | Colour |
| 0 | 0 | 1 | 1 | 1 | Black (off) |
| 0 | 1 | 0 | 1 | 1 | Green |
| 1 | 0 | 1 | 0 | 1 | Yellow |
| 1 | 1 | 1 | 1 | 0 | Red |

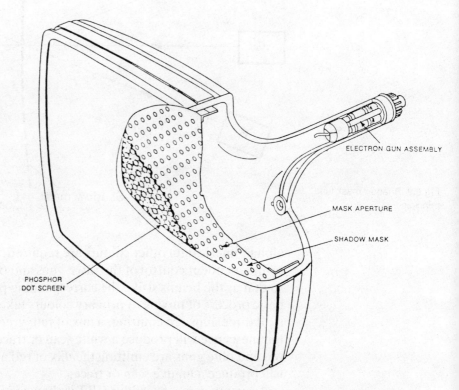

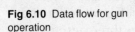

**Fig 6.9** Sectioned view of colour CRT

of which correspond to one of the colour guns, i.e., red, green and blue, as shown in Table 6.2.

The 'low' state outputs turn on the guns, and from Table 6.2 it will also be seen how simultaneous gun operation produces other colours from a mix of the primary colours. Figure 6.11 illustrates a typical weather data display together with associated alphanumeric data, namely ranges in nautical miles, and operating modes which in this case is WX signifying 'weather' mode.

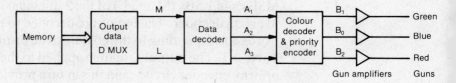

**Fig 6.10** Data flow for gun operation

114 Cathode ray tube displays

**Table 6.2** Colour decoding and priority encoding

| $B_1$ Green | Outputs to guns $B_0$ Blue | $B_2$ Red | Resulting colour display |
|---|---|---|---|
| 1 | 1 | 1 | Black (off) |
| 0 | 0 | 0 | White |
| 0 | 0 | 1 | Yellow |
| 0 | 1 | 1 | Red |
| 1 | 0 | 0 | Light blue |
| 1 | 0 | 1 | Green |

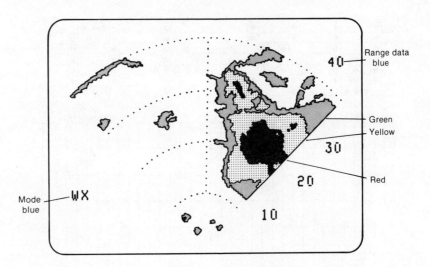

**Fig 6.11** Weather data display

## Alphanumeric Displays

The display of data in alphanumeric and symbolic form is extremely wide-ranging. For example, in a weather radar indicator it is usually only required for range information and indications of selected operating modes to be displayed, while in systems designed for comprehensive navigational and flight management functions a very much higher proportion of information must be 'written'. This is accomplished in a manner similar to that adopted for the display of weather data, but additional memory circuits, decoders and character and symbol generators are required. Raster scanning is also used, but where datum marks, arcs or other cursive symbols are to be displayed a stroke pulse method of scanning is adopted. The fundamental principle of generating alphanumeric displays is fundamentally the same as that employed in normal video display units. The position for each character on the screen is pre-determined and stored in a memory matrix, typically $5 \times 7$, and when the memory is addressed, the character is formed within a corresponding matrix of dots on the screen by video signal pulses produced as lines are scanned.

Figure 6.12 illustrates how, for example, the letters WX and the number 40 are formed. One line of dots is written at a time for the area in which the characters are to be displayed and so with a $5 \times 7$ matrix, seven image lines are needed to write complete characters and/or row of characters. Spacing is necessary between individual characters and between rows of characters, and so extra line 'blanking bits', e.g. three, are allocated to character display areas.

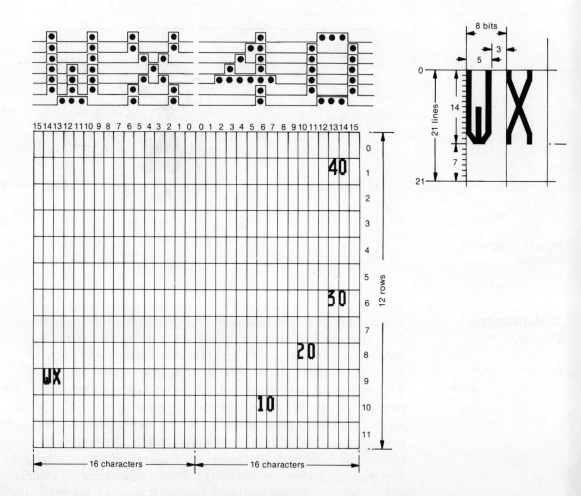

**Fig 6.12** Alphanumeric display

In the example of the weather radar indicator, the characters each have an allocation of 8 bits (5 for the characters and 3 for the space following) on each of 21 lines (14 for the character and 7 for the space below). The increase in character depth to 14 lines is derived from an alphanumeric address generator output that writes each line in a character twice during line scanning. The character format in this case permits the display of 12 rows each of 32 characters.

## Electronic Instrument Display Systems

The development of the colour CRT in association with that of data processing techniques, has made possible integrated displays of such a wide range of information that many individual traditional-type instruments, and caution and warning light systems, can virtually be dispensed with. At the present time, CRT or electronic instrument display systems are adopted in three public transport aircraft (see Figs 6.1 and 6.2) and also in some types of smaller aircraft operated for private and business executive purposes. There are three types of system, which are designated as: electronic flight instrument system (EFIS), engine indicating and crew alerting system (EICAS), and electronic centralized aircraft monitor system (ECAM). The EICAS and ECAM systems are not used jointly on any one aircraft type, and it is of interest to note that this results principally from the differing ideas of some aircraft manufacturers about the approach to such operating factors as: more efficient flight deck layouts and crew's controlling functions, the extent to which normal, alerting and warning information should be displayed, and in particular whether data relevant to engine operation requires to be displayed for the whole of a flight, or only at various phases. For example, EICAS was adopted for both the Boeing 757 and 767 aircraft so that engine parameters are displayed on their CRTs instead of by traditional instruments. These parameters, as well as other systems' parameters, are not necessarily always on display, but in the event of malfunctions occurring at any time the flight crew's attention is drawn to them by an automatic display of the data in appropriate colours. The ECAM system, as adopted for the Airbus A310, displays the operation of aircraft systems in a checklist and schematic form, and is based on the manufacturers' view that engine parameters need to be seen during the whole flight. Thus, for this aircraft separate traditional engine instruments have been retained.

EFIS is adopted for all three types of aircraft, and except for the display of speeds, certain display switching facilities and physical dimensions of display units, the design concept is broadly the same.

### EFIS

A complete EFIS installation in an aircraft is made up of left (Captain) and right (co-pilot) systems, each system in turn being composed of two display units (an attitude director indicator (ADI) and a horizontal situation indicator (HSI)), a control panel, a symbol generator and a remote light sensor. A third (centre) symbol generator is also incorporated so that drive signals from this generator may be switched to either the left or right display units in the event of failure of their corresponding generators. The signal switching is accomplished within the left and right generators using electromechanical relays powered from the aircraft's electrical system via pilot-controlled switches so that the switching function is not dependent on the operating condition of an affected symbol generator. The interface between units and intra-system signals is shown in Fig. 6.13.

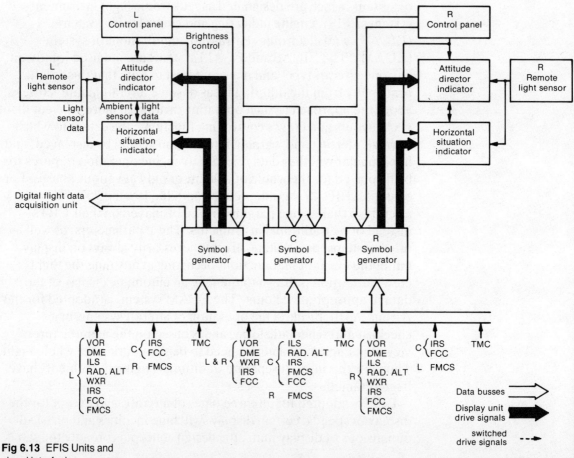

**Fig 6.13** EFIS Units and signal interfacing

*Display Units*

The display units consist of the following chassis-mounted elements: a low-voltage power supply, a high-voltage power supply, four circuit cards (video/monitor, convergence, deflection and interconnect) and a multi-colour CRT; all are contained within a protective cover. The principal differences between the ADI and HSI units is the CRT which, for obvious reasons, displays different levels of flight and navigation information. A simplified block diagram of a display unit is shown in Fig. 6.14.

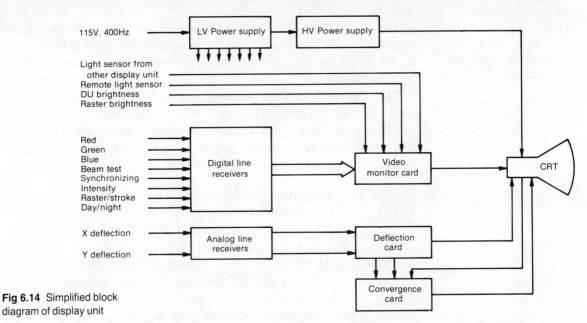

**Fig 6.14** Simplified block diagram of display unit

The power supply units provide the requisite levels of a.c. and d.c. power necessary for overall operation of the display units. The supplies are automatically regulated and are also monitored for undervoltage and overvoltage conditions.

The video/monitor card contains a video control microprocessor, video amplifiers and monitoring logic for the display unit. The main tasks of the processor and associated EPROM and RAM, are to calculate gain factors for the three video amplifiers (red, blue and green), and perform input sensor and display unit monitor functions. The input/output interface functions for the processor are provided by analog multiplexers, an A/D converter and a multiplying D/A converter.

The function of the convergence card is to take $X$ and $Y$ deflection signals and to develop drive signals for the three radial convergence (red, blue and green) coils and the one lateral convergence (blue) coil of the CRT. Voltage compensators monitor the deflection signals in order to establish on which part of the CRT screen the beams are located (right or left for the $X$ comparator and top or bottom for the $Y$ comparator).

Signals for the *X* and *Y* beam deflections for stroke and raster writing are provided by the deflection amplifier card. The amplifiers for both beams each consist of a two-stage preamplifier, and a power amplifier. The amplifiers use two supply inputs, 15 V d.c. and 28 V d.c.; the former is used for effecting most of the stroke writing, while the latter is used for repositioning and raster writing.

The interconnect card serves as the interface between the external connector of a display unit and the various cards. Digital line receivers for the signals supplied by the symbol generators are also located on this card.

The multi-colour CRT is a high-resolution shadow mask tube (see also Figs 6.8 and 6.9) using a delta configuration of the three electron guns. In a typical system, six colours are assigned for the display of the many symbols, failure annunciators, messages and other alphanumeric information, and are as follows:

*White*  Display of present situation information.
*Green*  Display of present situation information where contrast with white symbols are required, or for data having lower priority than white symbols.
*Magenta*  All 'fly to' information such as flight director commands, deviation pointers, active flight path lines.
*Cyan*  Sky shading on ADI and for low-priority information such as non-active flight plan map data.
*Yellow*  Ground shading on ADI, caution information display such as failure warning flags, limit and alert annunciations and fault messages.
*Red*  For display of heaviest precipitation levels as detected by the weather radar.

*Symbol Generators*

These provide the analog, discrete and digital signal interfaces to the aircraft systems, display units and control panel, and they perform symbol generation, system monitoring, power control and the main control functions of the EFIS. The interfacing between the card modules is shown in Fig. 6.15, and card functions are given in Table 6.3.

*Remote Light Sensor*

This is a photodiode device which responds to flight deck ambient light conditions and automatically adjusts the brightness of the electronic displays to a compatible level.

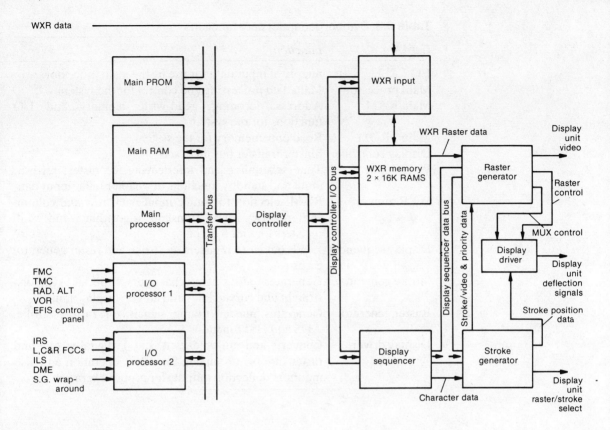

**Fig 6.15** Symbol generator and card interfacing

*Indicator Displays*

The ADI displays traditional pitch and roll attitude indications against a raster-scanned background, the upper half of which is in cyan and the lower half in yellow. Attitude information is provided by the inertial reference system. Also displayed are flight director commands, localizer and glide slope deviation, selected airspeed, ground speed, automatic flight control system and autothrottle modes, radio altitude and decision height.

Figure 6.16 illustrates a display representative of an automatically controlled approach to land situation together with the colours of the symbols and alphanumeric data produced via the EFIS control panels and symbol generators. The autoland status, pitch, roll-armed and engaged modes are selected on the automatic flight control system control panel and the decision height is selected on the EFIS control panels. Radio altitude is digitally displayed during an approach, and when the aircraft is between 2500 and 1000 ft above ground level. Below 1000 ft the display automatically changes

**Table 6.3** Symbol generator card functions

| Card | Function |
| --- | --- |
| I/O 1 & 2 | Supply of input data for use by the main processor |
| Main processor | Main data processing and control for the system |
| Main RAM | Address decoding, read/write memory and I/O functions for the system |
| Main PROM | Read only memory for the system |
| Display controller | Master transfer bus interface |
| WXR input | Time scheduling and interleaving for raster, refresh, input and standby functions of weather radar input data |
| WXR memory | RAM selection for single-input data, row and column shifters for rotate/translate algorithm, and shift registers for video output |
| Display sequencer | Loads data into registers on stroke and raster generator cards |
| Stroke generator | Generates all single characters, special symbols, straight and curved lines and arcs on display units |
| Raster generator | Generates master timing signals for raster, stroke, ADS and HSI functions |
| Display driver | Converts and multiplexes $X$ and $Y$ digital stroke and raster inputs into analog for driver operation and also monitors deflection outputs for proper operation |

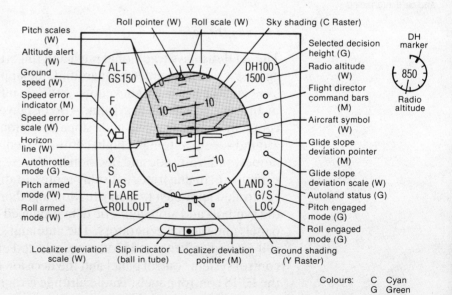

**Fig 6.16** Attitude Director Indicator

to a white circular scale calibrated in increments of 100 ft, and the selected decision height is then displayed as a magenta-coloured marker on the circular scale. The radio altitude also appears within the scale as a digital readout. As the aircraft descends, segments of the altitude scale are simultaneously erased so that the scale continuously diminishes in length in an anti-clockwise direction. At the selected decision height plus 50 ft, an aural alert chime sounds at an increasing rate until the decision height is reached. At the decision height, the circular scale changes from white to amber and the marker changes from magenta to amber; both the scale and marker also flash for several seconds. A reset button is provided on the EFIS control panel and when pressed it stops the flashing and causes the scale and marker to change from amber back to their normal colour.

If during the approach the aircraft deviates beyond the normal ILS glide slope and/or localizer limits (and when below 500 ft above ground level) the flight crew are alerted by the respective deviation pointers changing colour from white to amber; the pointers also start flashing. This alert condition ceases when the deviations return to within their normal limits.

The HSI presents a selectable, dynamic colour display of flight progress and plan view orientation. Four principal display modes may be selected on the EFIS control panel: MAP, PLAN, ILS and VOR. Figure 6.17 illustrates the normally used MAP mode display which in conjunction with the flight plan data programmed into a flight management computer (see page 180) displays information against a moving map background with all elements to a common scale. The symbol representing the aircraft is at the lower part of the display and an arc of the compass scale, or rose, covering 30 degrees on either side of the instantaneous track is at the upper part of the display. Heading information is supplied by the appropriate inertial reference system and the compass rose is automatically referenced to magnetic North (via a crew-operated MAG/TRUE selector switch) when between latitudes 73° N and 65° S and to true North when above these latitudes. When the selector switch is set at TRUE the compass rose is referenced to true North regardless of latitude.

The tuned VOR/DME stations, airports and their identification letters, and the flight plan entered into the flight management computer system (FMCS) are all correctly oriented with respect to the positions and track of the aircraft and to the range scale (nautical miles/inch) selected on the EFIS control panel. Weather radar returns (see page 113) may also be selected and displayed when required, at the same scale and orientation as the map.

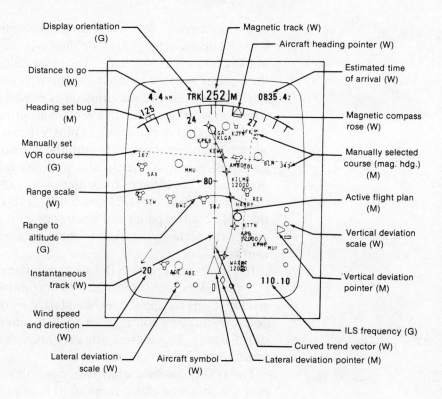

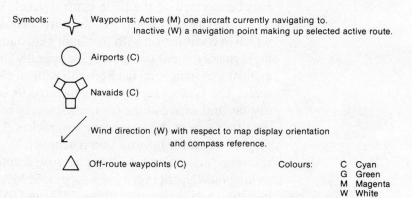

**Fig 6.17** Horizontal Situation Indicator in MAP mode

Indications of other data such as wind speed and direction, lateral and vertical deviations from the selected flight profile, distance to waypoint, etc., are also displayed.

The map display also provides two types of predictive information. One combines current ground speed and lateral acceleration into a prediction of the path over the ground to be followed over the next 30, 60 and 90 s. This is displayed by a curved track vector, and since a time cue is included the flight crew are able to judge distances directly in terms of time. The second prediction which is displayed by a range to altitude arc, shows where the aircraft will be when a selected target altitude is reached.

In the PLAN mode, a static map background with active route data oriented to true North is displayed in the lower part of the HSI display, together with the display of track and heading information (see Fig. 6.18). Any changes to the route are selected at the

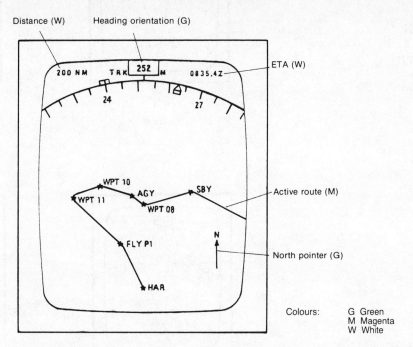

**Fig 6.18** Horizontal Situation Indicator in PLAN mode

keyboard of the display unit of the FMCS and appear on the HSI display so that they can be checked by the flight crew before they are entered into the flight management computer.

The VOR and ILS modes present a compass rose (either expanded or full) with heading orientation display as shown in Fig. 6.19. Selected range, wind information and system source annunciation are also displayed. If selected on the EFIS control panel, weather radar returns may also be displayed only when the mode selected presents an expanded compass rose.

*Failure Annunciation*

Failure of data signals from such systems as the ILS and radio altimeter are displayed on each ADI or HSI in the form of yellow flags 'painted' at specific matrix locations on the CRTs. In addition, fault messages may also be displayed, e.g. if the associated flight management computer and weather radar range disagree with the

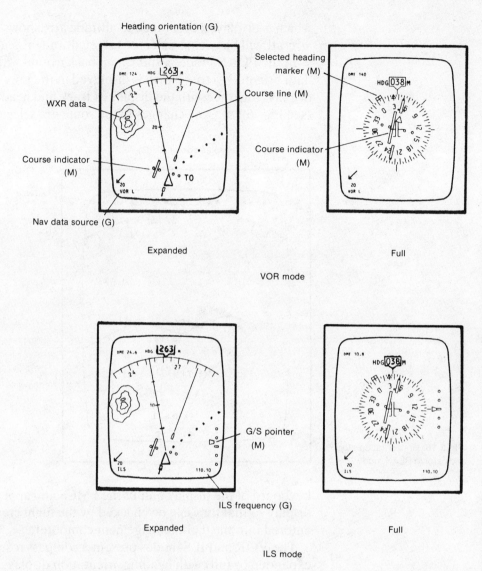

**Fig 6.19** VOR and ILS modes

control panel range data, the discrepancy message 'WXR/MAP RANGE DISAGREE' appears on the HSI.

## EICAS

The basic system comprises two display units, a control panel, and two computers supplied with analog and digital signals from engine and system sensors as shown in the schematic functional diagram of Fig. 6.20. Operating in conjunction with the system are discrete

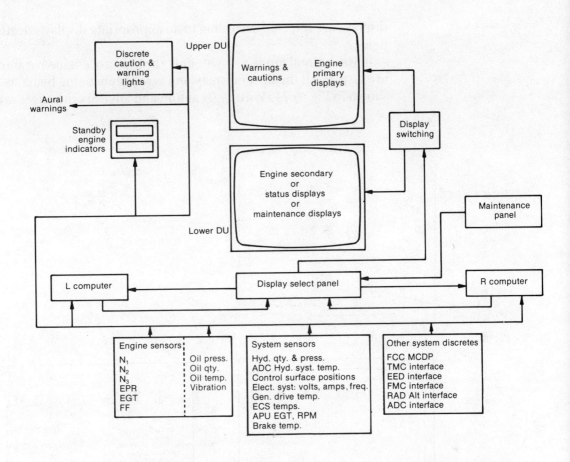

**Fig 6.20** Schematic functional diagram – EICAS

caution and warning lights, standby engine indicators and a remotely located panel for selecting maintenance data displays. The system provides the flight crew with information on primary engine parameters (full-time) and with secondary engine parameters and warning/caution/advisory alert messages (as required).

*Display Units*

These units provide a wide variety of information relevant to engine operation and operation of other automated systems and they utilize colour shadow mask CRTs and associated card modules whose functions are identical to those of the EFIS units described earlier. The units are mounted one above the other, the upper unit displaying the primary engine parameters (EPR, N1 and EGT) and warning and caution messages, while the lower unit displays secondary engine parameters (N2, N3, fuel flow, oil quantity, pressure and temperature, and engine vibration), status of non-engine systems, aircraft configuration and maintenance data. The

Microelectronics in Aircraft Systems 127

displays are selected according to an appropriate display selection mode.

In the normal mode in flight, only the primary engine parameters are displayed; the lower display unit screen remaining blank as shown in Fig. 6.21. Warning, caution and advisory messages are

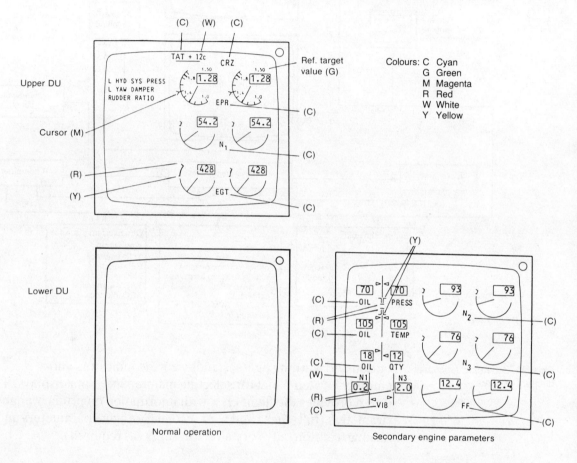

**Fig 6.21** EICAS Displays

displayed in red and yellow on the left-hand side of the upper display unit screen as conditions dictate. Abnormal secondary engine parameters are automatically displayed on the lower display unit.

As the various engine parameters change in value the corresponding indicator pointers rotate as in the case of traditional electromechanical indicators. With the exception of those indicated in Fig. 6.21, all elements of the displays are white under normal operating conditions. In the event of either N1 or EGT exceeding their normal values, both the corresponding pointer and digital readouts change from white to yellow, and to red, as the value reaches the limiting bands of the scale. In each case the highest value attained is displayed in white under the actual readout, and the accumulated exceedance time is stored in a non-volatile memory of the computer for subsequent readout during maintenance mode checks.

During the normal mode of operation all secondary engine parameters may be displayed on the blank lower display unit by pressing an 'ENG' select switch on the EICAS control panel. A second switch (STATUS) is provided on the control panel and when pressed it switches the lower display into a mode that displays the status of several systems (e.g. flight control surface positions) and also up to 16 status messages requiring flight crew awareness prior to take-off and in flight.

In order to ensure that all engine parameters can be displayed in the event that one display unit fails, the other unit can be operated in a compacted mode. The primary parameters are displayed as in the normal mode, while the secondary parameters are displayed in digital format on the same screen. Space is also available on the screen for a number of alert messages. The compacted mode format may also be displayed on one unit while the other is being used in either the status or maintenance mode.

In the event that multiple failures leave both display units or both computers inoperable, back-up displays by discrete caution and warning lights and by standby engine indicators are provided. These indicators utilize LEDs to supply EPR, N1, N3 and EGT information.

In order for maintenance engineers to carry out verification testing of systems and troubleshooting procedures, the lower display unit can be selected to display detailed data for several systems and up to 16 maintenance messages in what is termed the maintenance mode. Two examples of five possible formats are shown in Fig. 6.22, and the maintenance control panel on which selections are made is shown in Fig. 6.23. In addition, system failures which have occurred in flight and have been automatically recorded (auto event) in the computer memory, as also data on a suspect system problem entered in memory by the flight crew (manual event), can be called up for display by means of the 'event record' switch on the panel. A test switch is provided for initiating a self-test routine of EICAS.

## ECAM

The ECAM system comprises the units shown in Fig. 6.24. As far as display format is concerned, it differs significantly from EICAS in that it excludes analog presentation of engine parameters and it

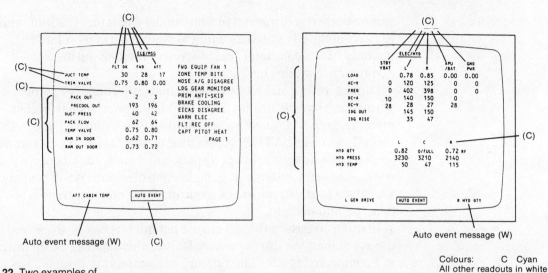

**Fig 6.22** Two examples of maintenance mode displays (lower display unit)

Colours: C Cyan
All other readouts in white

adopts the principle of mounting display units side-by-side so that the left-hand unit is dedicated to information on the system's status, warnings and corrective action in a sequenced checklist format, while the right-hand unit is dedicated to associated information in diagrammatic format.

There are four display modes, three of which are automatically selected and referred to as flight phase-related, advisory (mode and status) and failure-related modes. The fourth mode is manual and permits the selection of diagrams related to any one of twelve of the aircraft's systems for routine checking and also the selection of

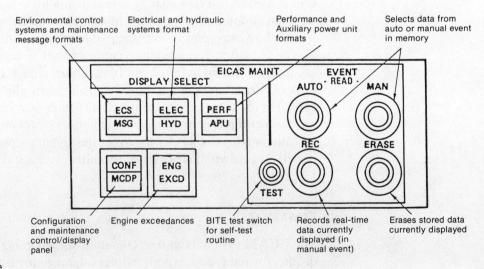

**Fig 6.23** Maintenance Control Panel

130   Cathode ray tube displays

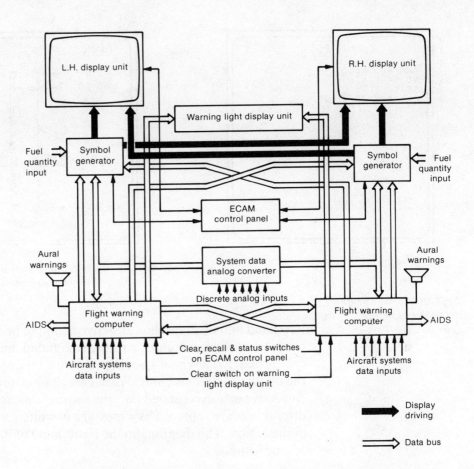

**Fig 6.24** ECAM System

status messages provided no warnings have been 'triggered' for display. The selections are made by means of illuminated push-button switches on the ECAM control panel.

In normal operation, the automatic flight phase-related mode is used and in this case the displays are appropriate to the current phase of aircraft operation, i.e. pre-flight, take-off, climb, cruise, descent, approach and after landing. An example of a pre-flight phase is shown in Fig. 6.25; the left-hand display unit displays an advisory memo mode and the right-hand unit displays a diagram of the aircraft fuselage, doors and arming of the escape slides deployment system.

The failure-related mode takes precedence over the two other automatic modes and the manual mode. An example of a display presentation associated with this mode is shown in Fig. 6.26. In this case, while taxying out for take-off, the temperature of the brake on the rear right wheel of the left main landing gear bogie has become excessive. A diagram of the wheel brake system is immediately displayed on the right-hand display unit and simultaneously the

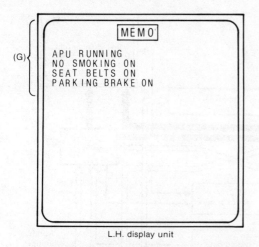
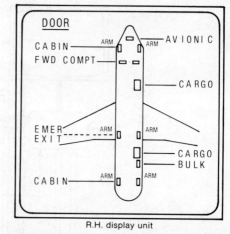

Examples: Doors locked: Door symbols green and name of door white
Doors unlocked: Door symbols and name of door amber

**Fig 6.25** Pre-flight phase-related mode display

left-hand unit displays the corrective action to be taken by the flight crew. In addition, an aural warning is sounded, and lights (placarded 'L/G Wheel') on the warning light display panel and on the overhead systems control panel are illuminated. As the corrective action is carried out, the instructions on the left-hand display unit are replaced by a message in white confirming the result of the action. The diagram on the right-hand unit is appropriately re-configured.

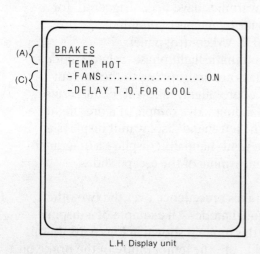

Colours: A Amber
C Cyan
G Green
Remainder of display white

**Fig 6.26** Failure-related mode display

132 Cathode ray tube displays

In the example considered, the warning relates to a single system and by convention such warnings are signified by underlining the system title displayed. In cases where a failure can affect other sub-systems, the title of the sub-system is shown boxed, as for instance in the display shown in Fig. 6.27. Warnings and the associated lights are cleared by means of CLEAR push-button switches on either the ECAM control panel or the warning light display panel.

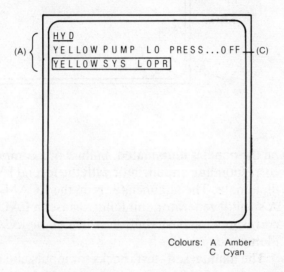

**Fig 6.27** Display of failure affecting a sub-system

Colours: A Amber
C Cyan

STATUS messages, which are also displayed on the left-hand display unit, provide the flight crew with an operational summary of the aircraft's condition, possible downgrading of autoland capability, and as far as possible, indications of the aircraft status following all failures except those that do not affect the flight. The contents and organization of an example display are shown in Fig. 6.28.

**System Testing**

Two self-test facilities are provided; automatic and manual. Each computer is equipped with a monitoring module which automatically checks data acquisition and processing modules, memories, and the internal power supply as soon as the aircraft's main power supply is applied to the system. A power-on test routine is also carried out for correct operation of the symbol generator units. During this test the display units remain blank.

In the event of failure of the data acquisition and processing modules or of the warning light display panel, an FWS warning light

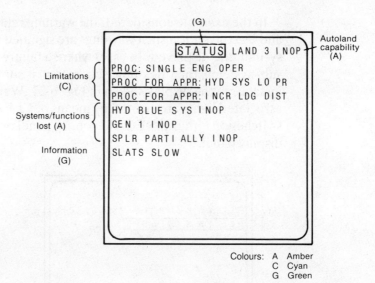

**Fig 6.28** Example of STATUS display

on the panel is illuminated. Failure of a computer causes a corresponding annunciator with the legend FWC FAULT to illuminate. The annunciator is on the ECAM control panel. A symbol generator unit failure causes a FAULT legend on the corresponding push-button switch of the ECAM control panel to illuminate.

The manual self-test checks for inputs and displays and is performed from a maintenance panel forming part of the ECAM system. When the INPUTS push-button switch is pressed a TEST legend is illuminated, and most of the inputs to each computer are checked for continuity. Any incorrect inputs appear in coded form on the left-hand display unit. The right-hand display unit presents a list of defective parameters at the system data analog converter. The diagrams of systems appear on the right-hand display unit with the legend TEST beside the system title, as each corresponding push-button switch on the ECAM control panel is pressed. Calibrated outputs from the data analog converter are also displayed. Any defective parameters are identified by a flag display.

A DISPLAYS push-button switch is provided on the maintenance panel and when pressed it initiates a check for correct operation of the symbol generator units and the optical qualities of the display units by means of a test pattern display.

# 7 Logic diagrams and interpretation

Prior to the currently wide-ranging applications of microelectronic circuit technology to aircraft systems being adopted, a study of the operation of electronic systems then in use necessitated a considerable amount of time having to be spent on tracing out some rather complex circuit diagrams. An example of one such diagram related to an automatic flight control system is shown in Fig. 7.1. At the time (c. 1948) this was a rather sophisticated system, and the author can recall how, during the course of instruction on the system, the instructor painstakingly went through each electron tube, and almost every resistor and capacitor, in order to cover amplifier operation. It was very time-consuming, but once the electrons had cleared, and signals were seen to be going in the right direction, the knowledge gained proved very useful, particularly in tracking down a fault, say, in a stage of a control channel. Invariably this would be caused by failure of a tube (not renowned for longevity), which could be simply replaced.

Systems using microelectronic circuits are by no means less complex in their design but, as far as the aircraft maintenance engineer is concerned, diagrams showing every single item on the lines of that in Fig. 7.1 are really no longer required. There are two principal reasons for this. First, all the major elements of circuits appropriate to a system perform known standard functions, and contain circuits which are so integrated that, after fabrication, they are totally inaccessible for any point-to-point testing. Second, with the application of logic gating concepts, the tracing of signal flows in microelectronic circuits has now become largely a matter of interpreting only schematic diagrams of interconnected 'blocks' and logic gate symbols. A lot of schematic diagrams may still be necessary, but 'hiding away' circuits in chips of silicon, and having a good understanding of element functions and of logic gate symbols, has made circuit tracing much easier.

As an aid to the interpretation of systems operation by means of schematic or *logic diagrams*, several examples of some representative aircraft systems will be described in the following

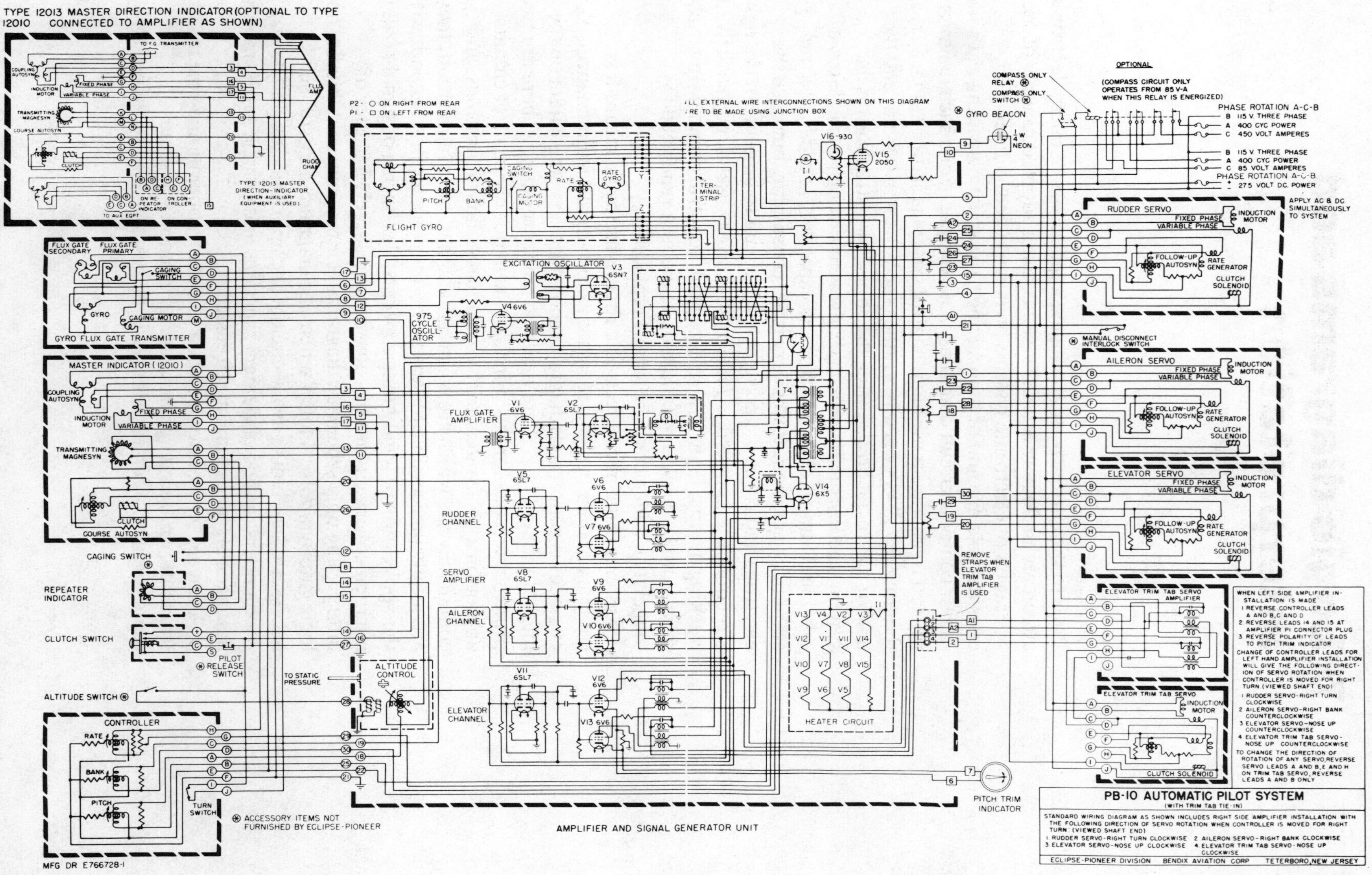

Fig 7.1 Circuit diagram of an AFCS using an electron tube amplifier (c. 1948)

sections. However, before we study these it will be found useful at this stage to expand on the subject of the logic circuit equations which were dealt with in Chapter 3 in order to understand their more practical applications.

## Equations and Logic Diagrams

All logic circuits are combinations of the three basic logic gates, and as reference to logic diagrams related to the functional elements of many current aircraft systems will show, these combinations can be very extensive. The writing of equations for these diagrams can therefore be an equally extensive operation requiring systematic expansion of the various gate functions within the combination, and grouping within the equations. This is a task for which systems and circuit designers are responsible, and no doubt, some of the practical maintenance engineers among the readers of this book will be thankful for that! However, it is customary for logic states, or for signal conditions expressed in Boolean equation form, to be detailed on the logic diagrams appropriate to a specific system as an aid to an understanding of its operation, and also to troubleshooting while it is in service. It is therefore of advantage to the maintenance engineer to have at least a working knowledge of how combinational logic circuit equations are developed from those of the three basic logic gates described in Chapter 3. With this in mind, some representative examples will now be considered.

In Fig. 7.2(a) an AND gate is shown providing an input to an OR

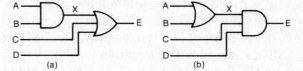

**Fig 7.2** Gate combinations. (a) OR matrix. (b) AND matrix

gate, this combination being referred to as an 'OR matrix' because the final gate is an OR. Four primary inputs are involved; the input $X$ is considered as a secondary input. The equation for the AND gate is, of course, $A \cdot B = X$, while the equation for the OR gate is $X + C + D = E$. The combined equation is therefore $A \cdot B + C + D = E$, which describes the structure of the logic diagram.

Figure 7.2(b) shows an OR gate feeding an AND gate, a combination known as an 'AND matrix' since the final gate is an AND. Similarly, there are still only four primary inputs involved, and the input $X$ is again considered only as a secondary input. The equation for the OR gate is $A + B = X$, while the equation for the

AND gate is $X \cdot C \cdot D = E$; thus, by combination the equation becomes $A + B \cdot C \cdot D = E$.

The combination of the foregoing equations has been done by a direct means which, though straightforward, can in some cases give a false impression of the overall diagram structure. For example, the equation for the AND matrix could be interpreted as '$E$ will only be present if we have $A$, or $B$ and $C$ and $D$'. To ensure that there is no misunderstanding of an equation, and of the signal functions in the various sections of a diagram, a system of grouping within parentheses must be adopted as in conventional algebraic equations. The equation for the AND matrix is therefore written as $(A + B) \cdot C \cdot D = E$. Parentheses are not used for an ANDed input to an OR or NOR logic symbol.

A more complex AND matrix is shown in Fig. 7.3. As before, the

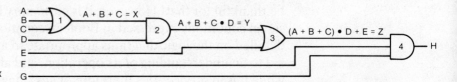

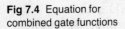

**Fig 7.3** Complex AND matrix

equation is developed by starting at the left of the diagram and finding the output of each logic gate. Gate 1 has three inputs and its equation is $A + B + C = X$, which is the secondary input to gate 2. This gate also has an input $D$, and so substituting for $X \cdot D = Y$ its equation becomes $A + B + C \cdot D = Y$. The inputs to gate 3 are $Y + E$, and so by substitution the output equation for gate 3 is $(A + B + C) \cdot D + E = Z$. Gate 4, from which in this example the overall output $(H)$ is to be provided, has three inputs $Z \cdot F \cdot G$. Thus, the complete equation for this combination is $[(A + B + C) \cdot D + E] \cdot F \cdot G = H$. It should be particularly noted how the identity and correct separation of quantities is maintained by parentheses and brackets as in conventional algebraic equations.

The development and application of Boolean equations applies equally to circuits which contain gates performing the NOT function, and an example is shown in Fig. 7.4. Gate 1 is a two-input

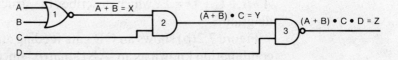

**Fig 7.4** Equation for combined gate functions

NOR and its equation is $\overline{A + B} = X$, which is the secondary input to AND gate 2. This gate also has an input $C$, and so its equation becomes $\overline{(A + B)} \cdot C = Y$. The inputs to NAND gate 3 are $Y$ and $D$, and so the complete equation for this combination is $(A + B) \cdot C \cdot D = Z$.

## Boolean Equations using Signal Functions

In the logic diagrams compiled for many aircraft systems, the logic states at the inputs and outputs of gates are given in Boolean equation form, but to relate the equations to the more realistic signal flow conditions, the letters $A$, $B$, $C$, etc., are substituted by abbreviated names of signal functions. For example, an equation term normally represented as A, may be designated 'FD CMD' which, in an actual system, signifies that the function is that of a 'Flight Director CoMmanD' signal. The significance of all abbreviations appropriate to any one system are given in the relevant maintenance manual (see also Appendix 6).

The same symbols and rules for grouping apply to signal function terms, and this may be seen from Fig. 7.5, which is an example of an

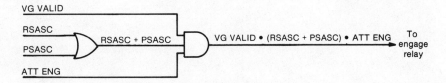

**Fig 7.5** Boolean equations using signal functions

AND matrix (see also Fig. 7.3) forming part of the engage logic in the computer of one type of automatic flight control system in current use. The full equation written at the output of the AND gate signifies that, for the engage relay to be enabled, three signal inputs must be present as follows:

1. a command signal to either the roll stability augmentation system (RSASC)
   **OR**
   the pitch stability augmentation system (PSASC);
   **AND**
2. a signal from the vertical gyroscope unit to indicate that it is powered and ready to perform its allotted function (VG VALID);
3. a signal to indicate that power is available for engaging (ATT ENG).

An example of a diagram using NOT signal functions is shown in Fig. 7.6.

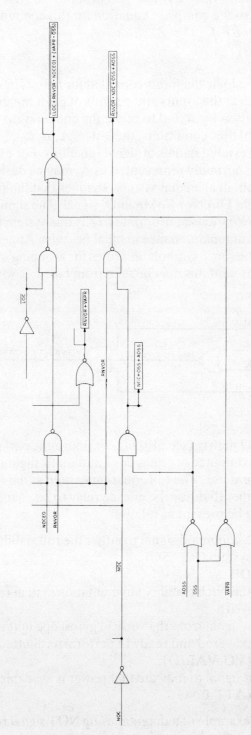

**Fig 7.6** Boolean equations using NOT signal functions

140   Logic diagrams and interpretation

# Examples of Practical Logic Diagrams

## Low Pressure Warning System

Figure 7.7 shows the logic of a system designed to give warning of low pressure in the pressurized cabin of an aircraft, and serves to illustrate the application of an inverter, a flip–flop and a driver.

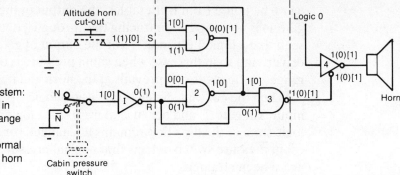

**Fig 7.7** Logic circuit of a low-pressure warning system:
1,0 Logic states when in normal pressure range
(1),(0) Logic states when pressure below normal
[1],[0] Logic states when horn cut-out is closed

The pressure switch senses cabin pressure and is adjusted to close under a pre-set low-pressure condition, and so provide a ground (logic 0) to operate the warning horn.

With the cabin pressure in its normal range, the switch is open, and a logic 1 is applied to the inverter input; this then becomes a logic 0 output and is applied to NAND gate 2 (the input at this gate being the equivalent R input of a flip–flop). Under these pressure conditions, the altitude horn cut-out switch is in the open position, and so a logic 1 is applied to NAND gate 1 (the input at this gate being the S input of the flip–flop). Thus, from the truth table for a flip–flop, logic 1 at S and logic 0 at R provides a logic 0 at the output of NAND gate 1 and therefore a logic 0 input to NAND gate 2. Gate 2 now has two logic 0 inputs, resulting in a logic 1 output which is applied to NAND gate 1 and its logic 0 output is thereby maintained. The logic 1 is also applied to NAND gate 3, and the second input to gate 3 is a logic 0; its output is thus logic 1 and is applied to the driver 4, the output of which is also a logic 1. The warning horn remains in a deactivated state.

If the cabin pressure should go below normal ($\overline{N}$) this will be sensed by the cabin pressure switch and its contacts will change over to provide a ground (logic 0) to the inverter. An inverted output, or logic 1, is now applied to the R input of the flip–flop together with the logic 1 input at S, and as is characteristic of such a device, with two logic 1 inputs applied to it the output logic state of its NAND gates 1 and 2 remains unchanged. However, it will be noted from the diagram that the logic 1 output from the inverter is also applied as an input to NAND gate 3, so that its output to the driver is changed to a logic 0, causing it to activate the horn which then gives warning of the low cabin pressure condition.

The warning horn may be deactivated by depressing the cut-out switch. This action changes the logic input at S of the flip–flop from a 1 to a 0, and since the input at R remains unchanged then, as may again be noted from truth tables, the output of the flip–flop (from gate 1) is now logic 1. In tracing this through gates 2 and 3 it can be seen that a logic 1 is produced at the output of gate 3, and of the driver, and so in this case when cabin pressure is below its normal range, an open-circuit prevails at the horn and it is thus silenced.

When the cut-out switch is released a logic 1 is again applied to input S at gate 1, and because there is still a logic 1 at R, the output logic state of the flip–flop remains unchanged (or 'SET') until the cabin pressure switch detects that normal pressure conditions have again been attained.

### AFCS Annunciator System

The logic circuit shown in Fig. 7.8 illustrates the application of OR, AND and NAND gates to an annunciator system associated with the approach mode of operation of an automatic flight control system (AFCS) roll channel.

As with any control system, the aircraft is controlled in its approach to an airport runway by signals beamed from localizer

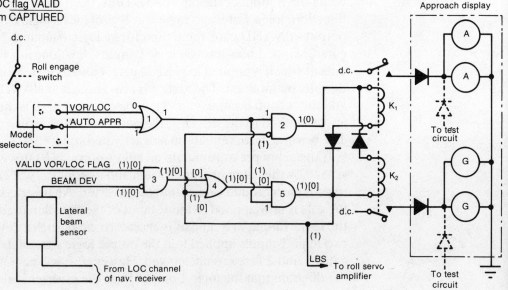

**Fig 7.8** Logic circuit of an approach display–localizer beam capture:
   1,0 Logic states on selection of AUTO APPR
(1),(0) Logic states valid VOR/LOC flag and beam captured
[1],[0] Logic states when VOR/LOC flag $\overline{\text{VALID}}$ and beam $\overline{\text{CAPTURED}}$

142   Logic diagrams and interpretation

(LOC) and glide slope (G/S) antennae of the Instrument Landing System (ILS), as indicated in Fig. 7.9. It is usual for some indication to be given to the flight crew that the automatic approach (AUTO APPR) mode of operation has been selected, and that 'capture', or interception, of the corresponding ILS beams has been accomplished. In the example system chosen, this is effected by sets of amber and green lights in an approach display unit, but for simplicity the operating and logic functions appropriate to the capture of the LOC beam only will be described.

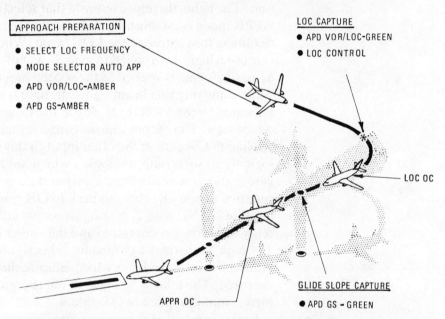

**Fig 7.9** Automatic approach path

Selection of the AUTO APPR mode is made on the control panel of the AFCS, and as may be seen from Fig. 7.8 a logic 1 input is applied to OR gate 1. Since the VOR/LOC position of the switch is open, then a logic 0 is applied as the other input to gate 1 and it produces a logic 1 state as its output, which is then applied to NAND gate 2. Sensing of any deviations from the LOC beam (the frequency of which is selected by the flight crew as part of approach preparation) is effected in the appropriate channel of the airborne navigation receiver, and for the operation of the circuit shown, two inputs from the receiver are required; the deviation signal, and a signal for the actuation of a VOR/LOC flag provided in the Course Indicator of the Flight Director System associated with the appropriate AFCS. The deviation signal is applied to a lateral beam sensor (LBS) within the roll control channel of the AFCS, and then as an input to NAND gate 3. The flag actuating signal is applied as the second input to gate 3.

Microelectronics in Aircraft Systems  143

When the values of the corresponding signals are below those required to actuate the VOR/LOC flag, and to initiate capture of the LOC beam, then the inputs to gate 3 are logic 0. Gate 3 assumes a logic 0 output state and applies this to OR gate 4, and because this gate now has two logic 0 inputs, its output applied to AND gate 5 is also logic 0. Together with the logic 1 input from OR gate 1, the output state of gate 5 is logic 0, and since this provides a ground for the logic 1 output state of NAND gate 2, then relay $K_1$ is energized to illuminate the amber lights of the approach display unit. The lights therefore indicate that selection of the AUTO APPR mode is established, and also that the system is armed in readiness for capture of the LOC beam. When the LBS senses beam capture a logic 1 is applied to gate 3. As noted earlier, the flag actuating signal is applied as the second input to gate 3 and is logic 1 state, signifying that beam signal strength is valid for control; this is indicated by the VOR/LOC flag in the Course Indicator being held out of view. The output state of gate 3 is therefore logic 1 and is applied to OR gate 4; the other input to this gate is momentarily logic 0 and so its output is logic 1 which, on being applied to AND gate 5 also produces a logic 1 output state at this gate. This output is supplied to relay $K_2$ and also back to OR gate 4 and to the inverted input of NAND gate 2, causing its output state to revert to logic 0; relay $K_1$ is now de-energized and the amber lights are extinguished. The logic 0 provides a ground for relay $K_2$ and so it is energized to illuminate the green lights which indicate that the beam has been captured. The relay is 'latched' in the energized condition by the logic 1 input supplied to OR gate 4.

Lateral control of the aircraft after beam capture, and ultimately onto the beam centre, is effected by also applying the logic 1 output state from gate 5 as the LBS signal to the amplifier which controls the application of power to the roll servomotor of the AFCS.

**Landing Gear Aural Warning System**

Figure 7.10 illustrates a logic circuit appropriate to a system the purpose of which is to operate a horn to warn the flight crew that the landing gear is not down and locked when the trailing edge flaps are set to the landing range, or when any engine thrust lever is set to the idle position. The switches A, B, C, D, E and F represent input sensors which are activated by the flaps, landing gear and thrust levers.

When the flaps are in the landing range (R) switches A and B are closed and so present logic 0 (ground potential) inputs to NAND gate 1; this in turn presents a logic 1 input to NAND gate 2. If the

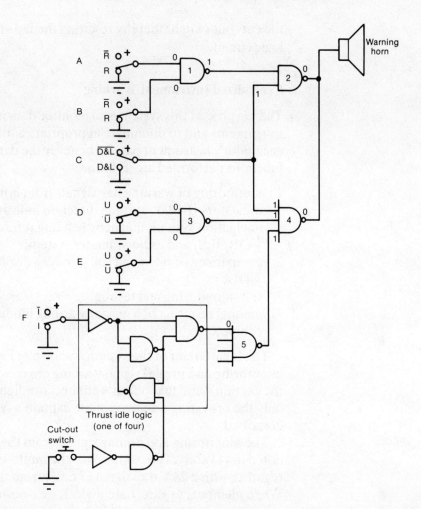

**Fig 7.10** Landing gear aural warning:
A and B: Trailing edge flaps setting
C: Landing gear
D and E: Trailing edge flaps up limit
F: Thrust lever idle

landing gear is not down and locked ($\overline{D \text{ and } L}$) a logic 1 from switch C will also be applied as a second input to gate 2, resulting in a logic 0 output to the horn, causing it to sound. The logic 1 is also applied as an input to NAND gate 4. Since the flaps are not fully up ($\overline{U}$), then switches D and E will apply logic 0 inputs to NAND gate 3, which produces a logic 1 output also to gate 4. When the thrust lever is in the idle position (switch F), its logic circuit produces a logic 0 input to NAND gate 5, and together with the logic 0 inputs from the thrust idle logic circuits related to the other engines, then the gate 5 output will be logic 1 and supplied as the third input to gate 4. Thus, the output of gate 4 remains logic 0 and the horn continues to sound. When the gear is fully extended to the down and locked position, the inputs to gates 2 and 4 from the switch C will be changed from logic 1 to logic 0 and the horn will therefore be silenced. The horn may also be silenced by depressing the cut-out switch, thereby resetting the flip–flop in the thrust idle logic circuit.

**Centralized Instrument Warning**

The purpose of this system is to monitor data inputs to the flight instruments and to illuminate appropriate annunciator lights on each pilot's instrument panel whenever the data is invalid. The functions performed are as follows:

1. monitoring of warning flag signals from horizontal situation indicators (HSIs), attitude direction indicators (ADIs), inertial navigation system, magnetic heading reference system, VOR/ILS, and radio altimeter system;
2. comparison of pitch and roll attitudes displayed in the HSIs and ADIs;
3. self-monitoring and testing;
4. manual resetting of a master warning module in the system's computer.

The logic circuit of the system is shown in Fig. 7.11. As far as flag monitoring and invalid signal warning are concerned the circuits to the captain's and first officer's annunciator lights are identical, so only the operating sequence of the captain's system will be described.

The monitoring and arming signals from the interfacing systems noted in (1) above, are supplied to the multi-input AND gate 1 together with a 28 V d.c. signal ($\overline{TEST}$) from the self-test control. When all inputs to gate 1 are logic 1, i.e., normal condition, the gate output is time delayed to prevent nuisance signals from triggering the logic. The logic 1 output is supplied to AND gate 2 together with input signals from the comparison warning logic which, as will be noted, are dependent on pitch and roll attitude, and heading signals via AND gates 4 and 5. In the normal condition, all inputs to gate 2 are logic 1, and so its output will be logic 1; this is fed to the S input of a flip–flop causing the $\overline{FAIL}$ output to be logic 1, and the $FAIL$ output logic 0. This prevents any output from AND gate 3, and so the instrument warning light remains extinguished.

When any input to the flag warning logic or the comparison warning logic fails, the inputs to gates 1, 4 and 5 will go logic 0 and so will their outputs. The output of gate 2 therefore goes logic 0 to reset the flip–flop so that the $Q$ or $FAIL$ output is now logic 1 and the $\overline{Q}$ or $\overline{FAIL}$ output is logic 0. The logic 1 output is supplied to AND gate 3 together with an input from a 1·5 Hz oscillator, and so

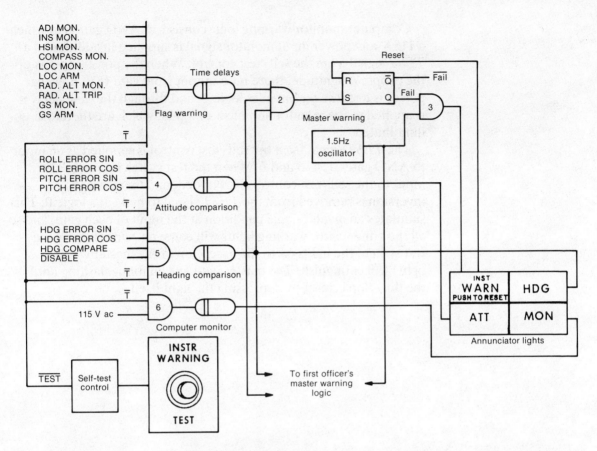

**Fig 7.11** Logic diagram of centralized instrument warning system

the gate 3 goes logic 1 at the positive peaks, causing the instrument warning light to flash on and off at the 1·5 Hz rate. When the failure is corrected, the lights continue to flash until the flip–flop is reset, i.e., until $\overline{Q}$ is logic 1. This is accomplished by depressing the warning light itself since it also serves as a push switch, and grounding the reset input of the flip–flop.

In the event that a difference exists between the pitch and roll attitudes indicated by the captain's and first officer's ADIs, the output of gate 4 in the comparison warning logic goes logic 0 and this causes the attitude annunciator warning lights on both pilot's panels to illuminate.

The logic controlling the heading warning annunciator light is identical to that controlling the attitude warning lights except that a 'heading compare disable' signal is fed as an input to AND gate 5 as well as heading error signals. This disables the heading comparison logic when either or both HSIs are displaying INS true heading instead of magnetic heading. When a difference of 6° in magnetic heading exists between the HSIs the heading annunciator light will be illuminated.

Microelectronics in Aircraft Systems 147

Computer monitor warning logic consists of AND gate 6 to which a 115 V a.c. power input monitor signal is supplied in addition to a $\overline{TEST}$ signal from the self-test control. When the power input is of the proper magnitude, there is no output from gate 6; if, however, power is lost, the output goes logic 1, and through the time delay it is applied to the monitor annunciator lights, which are therefore illuminated.

The $\overline{TEST}$ signal from the self-test control is supplied as an input to AND gates 1, 4, 5 and 6. When the test switch is pressed the input to the self-test control is grounded, and therefore the inversion is removed from the *TEST* signal to make it a logic 0. This simulates an invalid signal condition at the input of each gate, and so all the annunciator warning lights will come on. On releasing the test switch, the lights (with the exception of the master warning light) will extinguish. The master light will continue flashing until the flip–flop is reset by depressing the light itself.

# 8 Computers

Computers fall into two distinct categories, analog and digital, based on the methods by which they process data and perform appropriate calculations.

**Analog Computers**  Analog computers accept data as a continuously varying quantity, and create physical analogies of the variables in some other quantitative form such as a voltage or by the angular rotation of shafts or gearing. The magnitude of the variables are related by the physical laws on which input data are based.

When computers were first introduced into aircraft they were of the analog type and performed such functions as automatic flight control, and air data computation for the operation of servo-driven airspeed indicators and altimeters. The analog concept of measurement applied to them was, of course, nothing new since all conventional instruments indicate continuously varying quantities by creating analogies related to some or other physical law. For example, a basic form of airspeed indicator measures the speed of an aircraft in the analogous form of pressure differences and the mechanical deflection of a capsule type element, gearing and a pointer, and related by the $\frac{1}{2}\rho V^2$ law. Another example is a turbine gas temperature indicating system which measures temperature in the analogous form of an e.m.f. and the deflection of a moving coil and pointer, in accordance with the laws relating to thermoelectric effects.

Although analog computers provide for fast computation of continuously changing data, any system to which they are applied soon becomes unacceptably complicated. When the system is required to perform additional tasks, additional circuit elements and associated wiring are required. Furthermore, limitations in computer ability to store certain data, e.g., flight plan data, limits their usefulness.

## Digital Computers

A digital computer is capable of performing operations on data represented as a series of discrete impulses or bits arranged in coded form to represent numbers, alphanumeric characters and symbols. Figure 8.1 illustrates what is generally termed the organization

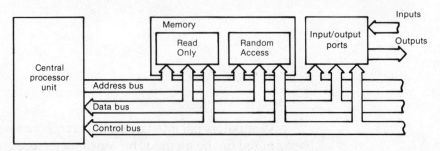

**Fig 8.1** Computer organization

(sometimes architecture) of the principal *hardware* elements of a digital computer. The central processor unit (CPU) executes the individual machine instructions which make up the computer program.

The binary coded format of the program consists of an *operation* code telling the computer what operation it is to start next, and an *operand* which is the data to be operated on. The program together with procedures and associated documentation form what is termed the *software*. The CPU contains a number of registers or temporary storage units which can each store a single byte or word, an arithmetic logic unit (ALU) which performs the binary arithmetic and basic logic functions associated with data manipulation, and a timing and control section for coordinating CPU internal operation so that fetching and execution of the instructions specified by a program is performed. The typical organization of a CPU is shown in Fig. 8.2.

Communication between the CPU and memory, and the input/output ports, is by means of a computer highway consisting of three separate buses; the data bus, address bus and control bus. The term 'bus' (derived from busbar which is a carry-all device) signifies a group of conductors carrying one bit per conductor, and is represented on diagrams by a broad arrow identified by function.

The data bus carries the data associated with a memory or input/output transfer, and the number of lines constituting the bus is the same as the number of bits (e.g. eight) in the CPU's word length. The bus is usually bidirectional, i.e., the CPU can write data to be read by a memory, or it can read data from the bus presented by the memory. Thus, data transfer between the two can be effected over a single set of data lines. All information transferred under program control travels on the data bus via the CPU.

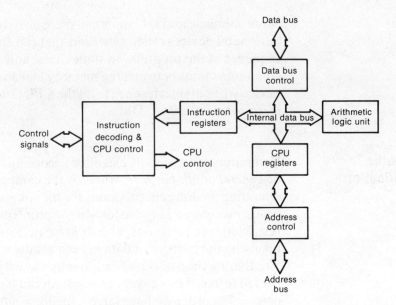

**Fig 8.2** Typical organization of a CPU

The address bus specifies the memory locations or input/output ports involved in a transfer. The number of bits constituting this bus has no direct relationship to the data bus word-length and depends on the operation being performed; e.g., at the beginning of an instruction cycle the CPU must supply the address of the next instruction in sequence to be fetched from the memory. Then, during the execution of the instruction, data may be required to be moved between the CPU and either the memory or an input/output port. If this is the case then the data memory address or input/output port address must be placed on the address bus by the CPU. Typically the bus contains 16 lines, and so gives the CPU the capability of addressing up to $2^{16}$ or 65 536 individual locations.

The control bus is also bidirectional; being made up of individual control lines for CPU memory and input/output control, it synchronizes the transfer of 'readout' and/or 'write-in' data along the data bus.

The memory consists of a number of storage locations for instruction words whose bit patterns define specific functions to be performed, and for data words to be used for carrying out the operations specified by the instruction words. Each memory word is given a numbered location called an *address* which itself is a binary word.

The input/output (I/O) ports form the interface between the computer and the source of input data and subsequent output data, and are generally under the control of the CPU. Special I/O instructions are used to transfer data into and out of the computer.

More sophisticated I/O units can recognize signals from extra peripheral devices called *interrupts* that can change the operating sequence of the program. In some cases, units permit direct communications between the memory and an external peripheral device without interference from the CPU; such a function is called *direct memory access* (DMA).

**Computer Classification**

Computers are generally classified according to function. They may be *general or all-purpose*, whereby the computer is capable of operating on different programs for the solution of a wide variety of problems; *special purpose* for solving problems related to specific or restricted type programs; or *hybrid* for performing specific tasks involving the transfer of data in both analog and digital forms. Another method of classification is by capacity, i.e., the volume of data in terms of bits a computer system can handle, and in descending order we have large-, medium- and small-size, mini- and microcomputers.

**Computer Languages**

In the same way that we humans communicate with each other through language, so a digital computer must use a language of one sort or another to carry out its functions. There is, however, a big difference between us and the computer in that when we are, say, given an instruction to do something, the understanding of our own language enables us to understand directly the instruction; apart from acting on it, no other conversion is required. This is not true for a computer, because when we want to give it an instruction a conversion from our language into the binary 1 and 0 coded language must first be carried out. The digital code is called *machine code*, and if instructions for the computer can be programmed directly in this code, the program is written in *machine language*, and overall it is called a *machine language program*.

The task of converting to machine code is usually delegated to the computer, which follows what is called an *assembler program* telling the computer what to do. The choice of instruction can be made by selecting a mnemonic which is an abbreviation of what the instruction does. This programming with mnemonic instructions is called *assembly language programming* because, after the sequence is written, it is fed into the assembler program which makes the conversion to machine code and assembles it into the memory in the proper order.

Such mnemonic codes are still not like human language, and so a *higher-level language* programming concept can be adopted in which instructions are written in a problem-orientated or procedure-orientated notation with each statement corresponding to several machine code instructions. The conversion of the statements to machine code is done by a more involved computer program called a *compiler*. The easier the programming is made by bringing the machine language closer to the human language, the more complex the computer program needed to convert the machine language statements in machine code. Once the conversion program is available, however, it can be used over and over again as necessary.

Details of some of the most common programming languages are given in Appendix 8.

## Microprocessors

Microprocessors are generally designed to perform a dedicated function and are built into equipment that will be used for some specific application. In terms of computing capability they are very powerful devices and extend the applications of computer techniques to many areas where minicomputers and microcomputers are not economically feasible. By the same token, they are also used as the main CPU of both types of computer, since they contain arithmetic logic and control sections. All of the standard logic functions, such as Boolean operations, counting and shifting, can be readily carried out by a microprocessor through programming, and as they also contain a limited form of I/O circuitry they may be directly interfaced with such devices as analog-to-digital and digital-to-analog converters, temperature and speed sensors, and LED matrix displays.

The photomicrograph of a microprocessor shown in Fig. 8.3 serves as a further example of the integrated circuit techniques referred to in Chapter 1. Figure 8.4 illustrates a fully packaged microprocessor which, together with the other two devices shown, can be used to form a complete and powerful microcomputing system.

## Data Transfer

The transfer of data between individual avionic systems is a necessary feature of aircraft operation. For example, under automatically controlled flight conditions, an automatic control system operates in conjunction with an inertial navigation system, air data computer, flight director system and radio navigation systems, and these involve the exchange of data to provide appropriate command signals to the control system.

When conventional techniques are used to interconnect all the

Fig 8.3 Micrograph of a microprocessor (courtesy Intel Corp.)

sensing devices, electronic control units, etc., the extent of the cabling required is considerable, particularly as individual wires must transfer signals dedicated to each of the parameters being measured. With the increasing application of digital computer-based systems, and in some of the current generation of aircraft there can be dozens of them, it has become necessary to adopt an alternative method by which the exchange of information can be effected by a network of single data busses (a *data highway*) within the aircraft. In other words, this is an adaptation of the highway concept adopted within digital computers themselves. Each data bus consists of a shielded and twisted pair of wires, and the voltage difference between them encodes a binary 0 or a 1. All outgoing encoded data from the computers are identified by an additional

**Fig 8.4** A microprocessor, memory and programmable input/output device (approximately 1½ times actual size) (courtesy Intel Corp.)

binary coded word called a *label*. The label takes up the first 8 bits of each word and is octal coded. Examples of outputs based on the ARINC 429 format developed for civil aircraft use are shown in Fig. 8.5.

The designation of labels to particular functions is arranged by the aircraft manufacturer in accordance with the tables given in the ARINC 429 specification. As separate bus systems are predictable for the different classes of aircraft system, the tables include some duplication of labels where it is known that the use of a common label on the same bus for two different purposes will occur. For example, label $315_B$ defines 'wind shear' for navigation systems, but the same label defines 'stabilizer position' for flight control systems.

The systems providing data outputs (referred to as transmitters) each have their own data bus connecting them to the 'receiver' systems in need of the data. Up to 19 receivers may be connected to a bus, and each receiver has a decoder which has to decode each word on the data bus in order to determine from a label which data word is of interest to a receiver. Data are transmitted in batches at a specified repetition rate along the appropriate busses, and at either high speed (100 kilobits per second) or low speed (12 to 14·5 kilobits per second) according to the frequency at which interfacing systems require an update of information.

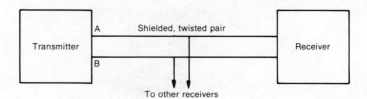

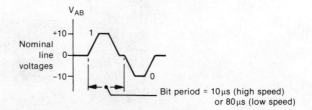

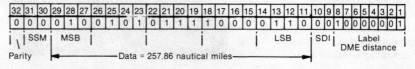

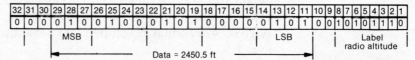

**Fig 8.5** Digital data transfer. SSM stands for Sign/Status Matrix and refers to plus, minus, north, south, etc., of BCD numeric data, the word type for AIM data, and the status of transmitter hardware. SDI stands for Source/Destination Identifier and is applied when specific words need to be directed to a specific system of a multi-system installation, or when the source system needs to be recognizable from the word content

## Memories

Since flip–flops and shift registers are devices capable of storing binary data (see Chapter 4), they may be considered as basic forms of memory device. This memory capability, however, is only of a temporary nature, and furthermore, it is limited to single bits of binary data. In many digital systems, computers in particular, it is essential to provide also for long-term and mass storage of information or programs made up of a much greater number of bits. Devices must therefore be built in to computers, and by virtue of having these capabilities can, in their own right, be classified as memories.

A memory consists of a large number of locations in each of which a small quantity of data can be stored, and these locations make it possible to write information into the memory or read information out of it. Each location has a unique address so that it can be accessed from outside the memory. The time that is needed to write one word into a memory or to read one out is referred to as the *access time* of the memory and is measured in nanoseconds ($1\,\text{ns} = 10^{-9}\,\text{s}$)

Accessing may be of two kinds: serial, in which the stored information is available for reading in a certain order, usually the same order in which it was put in (shift registers and magnetic bubble memories are examples using serial access), or random-access in which information can be taken out in any order.

**Capacity and Addressable Locations**

The capacity of a memory relates to 'bit storage' and is quoted in *kilobits* (K); the prefix kilo does not stand for 1000 as usual, but for $2^{10}$ or 1024. Thus $8K = 8 \times 1024 = 8192$ bits capacity.

The number of addressable locations in a memory is dependent on its number of input/output data lines, and is derived from the bit-storage capacity divided by the number of data lines. This is because each address location generally contains as many bits as it can pass through the data bus. If, for example, a 1K memory has only one data line, it will have 1024 separate addressable locations, but with four data lines it can only be addressed at 256 locations. The number of lines are decided by design and specified in the appropriate manufacturer's data sheets.

**Volatile and Non-Volatile Memory**

A *volatile* memory is one that will lose its stored data when the power supply is switched off, while in a *non-volatile* memory data is retained even though power is off.

**Random-Access Memory**

In a random-access memory (RAM) stored data at any location can be changed by 'writing in' new data at that location. It can therefore also be called a read/write memory. RAMs are either dynamic or static. In a dynamic RAM, data are represented by the presence or absence of electric charges which have to be 'refreshed' many times a second to prevent them (and the data) from leaking away. A static RAM has storage elements which act like bistable switches (e.g., flip–flops, diodes) so the data stored in them does not have to be refreshed; all that is required to keep the 'switches' on or off is a constant flow of current.

Information is stored in 'cells' which are arranged in an array or matrix of rows and columns, the number of which is governed by the storage capacity or 'organization' of the memory. Figure 8.6 illustrates the organization of an $8 \times 8$ bit matrix dynamic RAM. Each storage cell in this example consists of an individual MOS

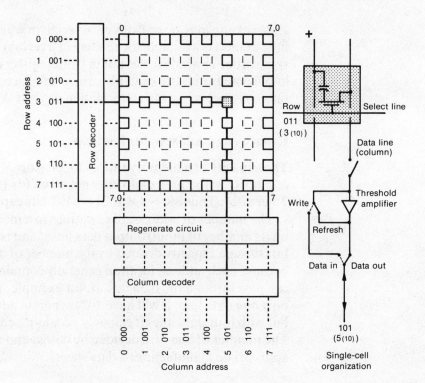

**Fig 8.6** Dynamic RAM operation

transistor and capacitor circuit as shown, and has a unique row and column address so that a particular cell can be selected by appropriate binary signals from row and column decoders. The purpose of the read/write control at the column decoder is to tell the decoder whether data are to be 'read out' or 'written in'. Cell locations start with 0 at the upper left of the matrix and end with 7 at the lower right; thus, in the example shown there are 64 bit locations.

To specify a particular memory cell location, three binary digits are needed to indicate the row addressed and another three to indicate the column. Assume, for example, that row address 3 (binary 011) and column address 5 (binary 101) are selected; then all the cell transistors in the row selection line will be turned on, but only the charge (binary 1) on the capacitor of the selected cell will be connected to the data line which is to transfer one bit of information, i.e., column 5. Because a cell capacitor loses charge by being 'read', or by leakage, there is a possibility of data being lost. This is prevented by a threshold amplifier in the data lines, the supply from which periodically (e.g., once every 2 ms) regenerates or refreshes the charge.

158 Computers

## Read-Only Memory

The main feature of a read-only memory (ROM) is that the binary information it contains is permanently stored in it. The data, which can be accessed in random fashion, are written in at the time of manufacture, and so the specific program contents cannot usually be changed afterwards. Although a ROM has only one decoder, the general organization is basically the same as a RAM, and in the organization of some digital systems the two types of memory are used together. The operation of a ROM constructed with a 'one of eight' decoder and a diode matrix is shown in Fig. 8.7. The decoder

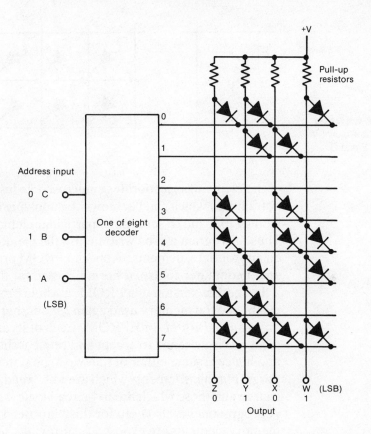

**Fig 8.7** Read-only memory (ROM)

accepts a 3-bit address input word, and is so called because in recognizing the word it will enable only one of the eight outputs. For example, if the word is 011 (i.e., $3_{(10)}$) the number 3 output line will go logic 0, while all the other output lines will be logic 1. The cathode ends of the diodes connected to line 3 will also go logic 0, causing the diodes to conduct through their associated pull-up resistors, and forcing lines X and Z to be logic 0. Since all other decoder outputs are logic 1 the other diodes in the matrix are cut

Microelectronics in Aircraft Systems 159

off, and so lines W and Y are logic 1. Thus, at address location 010, the word 0101 is stored and would be read out from the memory.

The contents of a ROM cannot usually be changed after manufacture, and this is due to the special masking technique adopted in programming the chip. However, to meet the needs of a user who may wish to do his own full programming, or alter an existing program, other techniques are adopted to produce *programmable ROMs*, or PROMs for short. In one version of a PROM, the chip is supplied ready-made, but with every one of its memory elements linked to read logic 1, as shown in Fig. 8.8. Each

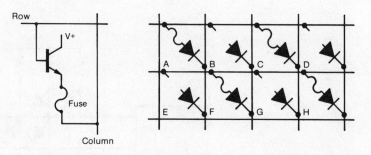

**Fig 8.8** PROM with fusible link. The fuses are 'blown' in elements B, D, E and G

of the elements incorporates a microscopic fuse (typically polysilicon) which can be 'blown' by applying an appropriate voltage to it and so setting the particular element to read logic 0. Thus, a program can be written into the memory by blowing fuses on appropriate elements. A special PROM programmer operating under computer control is normally used for this purpose.

In another version of a PROM, a whole program can be 'wiped clean' and the memory used again for another program; this is called an *erasable PROM*, or EPROM. Each of its array of memory elements is designed to accept and retain a charge of electrons, and hence assume either of the two logic states needed for programming. Elements which are to be read logic 1 are given a charge and those which are to be read logic 0 are left uncharged. A programmer is also used for this purpose. To erase a program the memory elements are exposed to ultraviolet light radiated through a 'window' in the cover of the IC pack (Fig. 8.9), causing the electron charge to leak away.

A third version of a PROM is one which can be electrically programmed and erased, known as an *electrically alterable* ROM or EAROM. In contrast to an EPROM it provides control over erasure. With an EPROM the contents of particular memory elements cannot be rewritten without first erasing the whole memory, but with an EAROM the contents of individual elements

**Fig 8.9** A UV EPROM (courtesy Intel Corp.)

can be erased and changed without disturbing the data in the other elements.

**Bubble Memory**

A bubble memory (Fig. 8.10) is a serial access device in which signal bits to be stored are transferred as minute cylindrical magnetized domains or bubbles (about 5 $\mu$m diameter) in a sequential manner in a thin film of a certain magnetic crystalline material. It provides for reading, writing and erasing of data, and is applied to the computers used in a number of current avionic systems.

The magnetic crystalline material (such as yttrium–iron garnet) is grown epitaxially on a substrate of non-magnetic crystalline material called gadolinium–gallium garnet. The chip is sandwiched between two flat permanent magnets that provide similar main fields pointing in the opposite direction to that of the bubbles, so that they form inverted spots in the main field. The bubbles are generated by passing pulses of current in one direction through aluminium conductor loops over an insulating film of silicon oxide on top of the epitaxial layer. A bubble under a loop can be erased by passing current pulses through the loop in the opposite direction.

The bubbles are stored along definite paths in a pattern of nickel–iron strips (e.g., Permalloy) over the silicon oxide film, by passing alternating current through two sets of coils disposed around

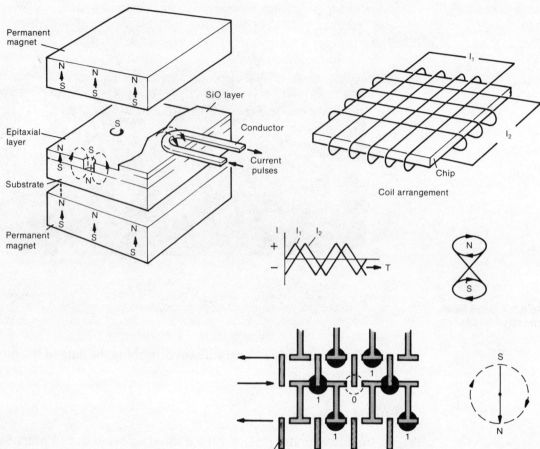

**Fig 8.10** Bubble memory

the chip at right-angles to each other. The current generates weak magnetic fields which cause the effective main field of the permanent magnets to rotate clockwise and set up a conical or 'wobbling' movement. As the wobbling of the main field takes place, it makes minute bar magnets of the segments of the nickel–iron strips that are most nearly aligned with the main field. Each bubble moves to seek the nearest north pole of a segment, and so the paths provided by the nickel–iron strips keep the bubbles spaced out (about 20 $\mu$m) one bubble per segment along the path. The presence of a bubble stores a binary 1, and the absence of a bubble stores a binary 0. Each time the field wobbles around for one revolution, all the bubbles advance one step and in opposite directions on alternate rows. The advancing or shifting frequency is determined by the frequency at which field rotation and wobbling are generated. The bubbles stay where they are when power is switched off, and so the device is non-volatile.

**Data Conversion Devices**

Many of the parameters associated with aircraft systems operation are measured in analog form. To process the appropriate signal data by means of digital computing techniques, additional hardware elements are therefore required to perform the conversions from analog-to-digital and from digital-to-analog. These are referred to as A/D and D/A converters and they respectively perform the functions of encoders and decoders, as shown graphically in Fig. 8.11.

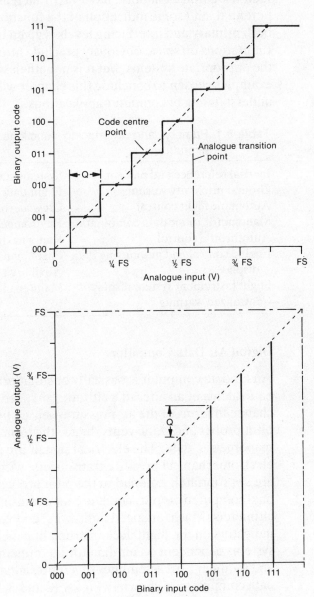

Fig 8.11 Transfer functions of ideal A/D and D/A converters: FS = Full scale; Q = Analog step magnitude or quantum

**Computer Applications**

One of the first digital computer-based systems to be used in commercial aircraft operations was the inertial navigation system. This paved the way for area-navigation systems to perform lateral and vertical navigation functions, which in conjunction with those of automatic control, flight director, and full flight-regime autothrottle systems, provided for more efficient flight operations. The continuing technological development of digital computers and associated hardware, together with such other factors as more cost-effective equipment installations and improved reliability and fault-isolation capability, have vastly increased their applications potential, and some indication of the functions they now perform at both primary and interfacing levels is given in Table 8.1. Limitations on space obviously preclude attempts at describing all of the appropriate systems, but it is nonetheless hoped that the examples chosen to conclude this chapter will help in the understanding of computer applications.

**Table 8.1** Primary and interfacing functions of digital computers in aircraft

| | |
|---|---|
| Inertial reference and navigation | Systems status display |
| Ground proximity warning | Flight data recording and acquisition |
| Automatic flight control | Crew alerting |
| Manometric or air data computation | Environmental control |
| Autothrottle control | Radio navigation and ATC |
| Engine control and monitoring data display | Flight management and performance |
| | Auxiliary power unit control |
| Flight instrument system display | Maintenance control and data display |
| Centralized warning | |

### Digital Air Data Computer

An air data computer is basically one that processes electrical signal equivalents of an aircraft's altitude, airspeed and rate of altitude change in terms of the air pressures sensed by externally mounted pitot probes and static vents; hence the term air, or sometimes manometric, data. The electrical signals are produced by electromechanical pressure transducers, within the computer, which are appropriately coupled to the pitot probes and static vents. The final computed output signals are then transmitted to servo-operated altimeters, Mach/airspeed indicators, and vertical speed indicators on the flight deck instrument panel, and also to all other systems dependent on air data for their operation. The organization and modular card assemblies of a representative type of digital air data computer are shown in Figs 8.12 and 8.13.

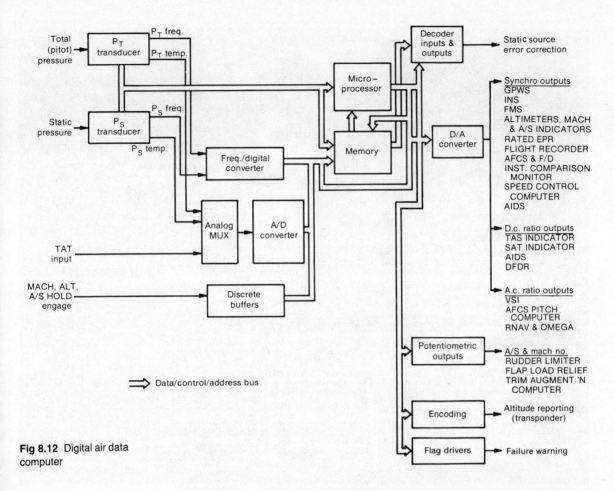

**Fig 8.12** Digital air data computer

In addition to the primary analog pitot and static pressure inputs, analog signals corresponding to the total air temperature sensed by an externally mounted probe are supplied as an input. These signals are required for the computation of true airspeed. Conversion of the inputs to serial digital format is accomplished by analog multiplexing, frequency/digital, and A/D conversions techniques. The converted data, also in serial format, are supplied to a microprocessor and memory for generation of the output signals required by the systems and equipment which interface with the computer. As will be noted from the diagram, there are four principal output stages: (i) D/A conversion providing analog outputs for the operation of synchros, and for d.c. and a.c. ratios required by the indicators and systems noted; (ii) outputs equivalent to a potentiometer output and corresponding to airspeed and Mach number: (iii) encoding of altitude data for a transponder system (see Chapter 9); and (iv) flag drivers for the operation of indicator flags to warn of output data failure.

**Fig 8.13** Card assemblies of a digital air data computer (courtesy Lockheed)

In connection with the sensing of both pitot and static pressures, the locations of pitot probes and static vents must be carefully chosen to minimize any error in the indications of airspeed and altitude. The error, which is called pressure or position error (PE), is related principally to static pressure sensing, and is more closely defined as the amount by which the local static pressure at a given point in the flow field differs from the free-stream static pressure. In addition to location, PE can also be influenced by other factors, e.g., by changes in airflow and aircraft attitude when flaps and other lift augmentation devices are deployed. The actual PE due to a chosen location varies between each type of aircraft and is determined during the initial flight-handling trials of a prototype; it is presented in tabular or graphical form enabling corrections under various operating conditions to be applied. In a digital air data computer corrections are programmed into the microprocessor and memory so that in operation they are automatically applied to the measured airspeed and altitude data via a static source error-correction decoder.

**Performance and Failure Assessment Monitor (PAFAM) System**

This system is also one which uses a digital computer and a colour CRT display, its purpose being to operate in conjunction with an automatic flight guidance system (AFGS) to provide a flight crew with a prediction of the quality of an automatic approach and landing manoeuvre being carried out in low visibility. It monitors aircraft attitude, heading, and performance of the AFGS and makes a continual assessment of whether or not a successful automatic landing will result. In the event that the progress of the manoeuvre is unsuccessful, a 'TAKEOVER' command is displayed; if the aircraft is being flown manually with commands from the flight director system, and the approach path is unacceptable, the legend NO TRACK is displayed. A block diagram of the system is shown in Fig. 8.14.

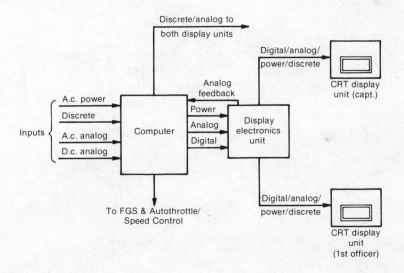

**Fig 8.14** Performance and failure assessment monitor system

The computer accepts electrical input signals from those sensors and sub-systems necessary for proper operation of the AFGS and auto throttle/speed control system. Electrical power is applied when the AFGS LAND ARM mode or flight director ILS modes of operation are selected, and the system is automatically switched to its operational condition when the ILS localizer and glideslope are being tracked.

The signal inputs to the computer are a.c. and d.c. analog and are multiplexed into an A/D converter which is under programmed memory control by one of two control processors in the computer; this processor performs most of the landing performance and prediction computations. Discrete signal inputs are multiplexed directly into the second processor which provides display drive commands, landing system failure assessment, and controls signals for discrete outputs. Interconnection between the two processors is through two 18-bit storage registers.

Analog signals from the computer are applied to the display electronics unit, and they provide commands for blanking out a portion of two raster-scanned CRT display units (one for each pilot) as well as commands which determine the location of desired characters in the display. The location of a display unit is shown in Fig. 8.15; the viewing area of the CRT is 38 mm × 76 mm. Discrete

**Fig 8.15** Location of PAFAM CRT display unit (Captain's position). Display unit

signal outputs are supplied to the AFGS and auto throttle/speed control system. The digital signal outputs from the computer are applied to timing and logic circuits in the display electronics unit for the development of analog character signals via fixed memory circuits in a symbol generator. The character signals are amplified by horizontal and vertical summing amplifiers, and then fed to deflection amplifier and blanking circuits so that desired symbols and words are 'painted' on the CRT screen. A colour control logic circuit supplies the CRT with a command signal which varies the level of a high-voltage supply so as to vary the colour. As the voltage is increased in selected steps, the colour of the character or raster being generated changes from red to red-orange, to amber, to yellow, and finally to green.

*Operation*

When the LAND or ILS mode logic is available from the AFGS computers, a TEST mode display first comes into view (Fig. 8.16(a)). Then, after a very short time period, a raster pattern is displayed on the CRTs representing the airport runway over which is superimposed a cross depicting the predicted touchdown point, as shown in Fig. 8.16(b). The runway symbol has a yellow border within which there is a green area representing the centre

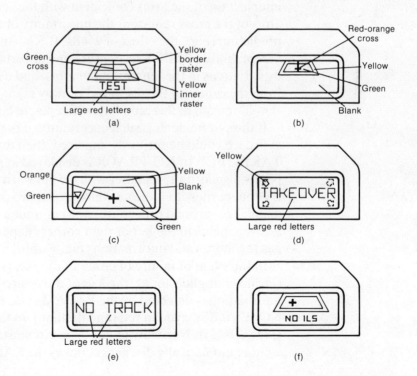

**Fig 8.16** PAFAM system displays: (a) Test pattern mode. (b) Beginning of approach. (c) Touchdown. (d) Takeover. (e) No track. (f) No ILS

Microelectronics in Aircraft Systems 169

touchdown zone. The area corresponds approximately to a zone ±18 m (60 ft) laterally and ±300 m (1000 ft) longitudinally about the nominal or ideal touchdown point, which is on the centreline and approximately 150 m (500 ft) beyond the glideslope transmitter location. The horizontal line closest to the bottom of the display corresponds to the runway threshold.

The symbol 'expands' by moving downwards as a function of computed range in a manner corresponding to the same rate of expansion that would be apparent if the real runway were visible out the flight deck windscreens. As the aircraft crosses the runway threshold, the bottom line of the display moves out of view, and on the basis of range computation, the lower edge of the green area reaches the bottom of the display when the aircraft passes over the nominal touchdown point. The green area continues to move down until touchdown. At touchdown, a green downward-pointed triangle is displayed on the CRT screen (Fig. 8.16(c)) and the display then remains static for about three seconds, after which the whole system is de-energized.

The orange-coloured cross symbol is positioned to show the predicted touchdown point on the runway as determined by simplified dynamic models of the aircraft/AFGS combination operated in an accelerated time scale. In an ideal landing situation, the cross will be superimposed over the touchdown area, with the intersection of the arms coincident with the centre of the area. The arms of the cross represent the uncertainty of the basic touchdown prediction (e.g., the effect of winds, ILS beam anomalies, and acceptability of AFGS responses), the uncertainty being reflected by variations in the lengths of the arms, and deflection of the cross from the centre of the displayed runway area. All parts of the cross should remain in this area for an acceptable landing.

If the system detects such uncertainty of performance that an approach could be seriously impaired, then the message TAKEOVER or NO TRACK is displayed in red letters on each pilot's display unit, together with a yellow arrow symbol in one of the four corners to indicate which of the two pilots should perform a manual go-around procedure, or a continued manual landing. The arrow appears in the left or right corners depending on which pilot has the more valid information (Fig. 8.16(d)).

In the event of failure of either of the ILS ground transmitters (localizer or glideslope), the loss of the desired guidance signal information is detected by the PAFAM system and the message NO ILS is displayed in red (and flashing) on the indicator units (Fig. 8.16(f)). Normally under these circumstances the AFGS would automatically disengage, but as the PAFAM system uses other references to supply equivalent ILS deviation for assessing performance, then if no deterioration in performance is detected, disengagement of the AFGS is inhibited for up to 5 s. If, during this interval, ILS signals are restored, the PAFAM system reverts to the normal display mode and the landing continues uninterrupted. If the signals are not restored, or if the performance is affected, the AFGS would then disengage and the NO ILS message in the display units would automatically change to TAKEOVER.

An automatic self-testing function is built into the system and checks the whole of its operation twice each second. Any malfunction causes latching-type annunciators to trip, and gross distortions or blanking out of symbols on the display units.

**Performance Advisory and Flight Management Systems**

Systems designed in various forms to carry out performance advisory or comprehensive flight management functions are now an essential feature of a number of types of commercial transport aircraft, their development having stemmed from the need to ensure the most efficient use of fuel, the need to reduce workload and the need to reduce operating costs overall. Fuel usage and other economic factors associated with aircraft operations have always been ones attracting the attention of the manufacturing and operating sectors of the industry, but in about the early 1970s when certain of the oil-producing states were creating sharp increases in the costs of crude oil and for political reasons were imposing oil embargos against some Western nations, the industry was forced to pay even greater attention to the above factors. As a result, many research and development programmes were instituted and were centred on the fuel efficiency of engines, improvements in $L/D$ ratios of airframes (e.g. by such means as use of supercritical wing sections) and on reductions in structural weight by use of composite materials.

Computer technology, although limited at the time in its application to aircraft systems, was nevertheless more advanced in overall concepts, and so by the production of software, which took into account the many operational variables, computers offered an additional and quicker route to the attainment of economic flight operations by performing automatic adjustments to relevant control systems. In conjunction with developments in the areas noted earlier, a system of computerized flight management has currently become the 'élite' of avionic systems and it is probably not unfair to say that it is the most developed, besides being the one most readily retro-fitted to aircraft.

FMS development has, of course, resulted in a number of variations on the original theme of controlling engine power and flight operations consistent with the most efficient use of fuel at all times and, consequently, a variety of system designations has been applied by the manufacturers of systems. These designations, some of which are interchangeable, while others indicate distinctly different capabilities, are given in Table 8.2. Although Table 8.2

**Table 8.2** System designations

| System | Function |
| --- | --- |
| Performance advisory system (PAS) | Advises of best altitude and speed to fly at to save fuel. Flight crew have to transfer values into automatic flight control system and throttle settings |
| Performance data computer system (PDCS) | Similar to PAS but typically linked to provide automatic pitch and throttle control |
| Performance management system (PMS) | Similar to PDCS but with additional lateral navigation capability |
| Automatic performance management system (APMS) | Similar to PAS |
| Flight management computer system (FMCS) | Full performance and navigation capabilities, flight planning and operation in a three-dimensional capacity |
| Flight management system (FMS) | Similar to FMCS |

has not been compiled to a rigid scale of evolution, it does provide some indication of development of system functions which may be advisory only, or a combination of advisory and control.

In performing an advisory function a system merely advises the flight crew of the optimum settings of various control parameters, such as engine pressure ratio (EPR) and climb rate under varying flight conditions, in order to achieve the most economical use of the available fuel. Such systems require adjustments of controls on the part of the flight crew if they are to be utilized to maximum advantage. Examples of advisory systems are the PAS and PDC systems noted in Table 8.2.

A system performing a combined function is one in which the sensing computer and display units are interfaced with an auto-throttle control system and pitch channel of an automatic flight control system; thus, in removing the flight crew from the control loop, an integrated automatic FMS is formed to provide greater precision of engine power and vertical flight path control.

Early forms of flight management systems, whether purely advisory or combined function, were limited to supervising control parameters affecting the vertical flight path. In order to ensure maximum fuel economy it is, however, also necessary to integrate this optimized flight path management with the lateral flight path; in other words, a system must also be provided with a navigation capability. This requires interfacing the computer with such navigation systems as Doppler, inertial reference system, DME and VOR. The inputs from these systems permit continuous monitoring of an aircraft's track in relation to a flight plan which may be pre-stored in the computer memory and an immediate identification of deviations. Furthermore, it allows flight plan variants to be constructed and evaluated. It is thus apparent that by combining these inputs with those controlling the vertical flight path parameters mentioned earlier, an FMS can integrate the functions of navigation, performance management, flight planning and three-dimensional guidance and control along a pre-planned flight path.

*Inputs*

In addition to changing data inputs from such systems as those mentioned above, an FMS system also requires data bases for storing bulk navigation data, and the characteristics of an aircraft and its engines, in order that the system will operate in a full three-dimensional capacity. The navigation data base is capable of storing the necessary flight environmental data associated with a typical airline's entire route structure, including pertinent navigation aids and waypoints, airports and runways, published terminal area procedures, etc. The memory bank also contains flight profile data for a variety of situation modes, such as take-off, climb, cruise, descent, holding, go-around and 'engine-out'. The cruise mode is also sub-divided into sub-mode variants such as economy, long-range, manual and thrust-limited. The integration of all the foregoing data, plus other variable inputs such as wind speeds and air traffic control clearances, permit the automatic generation or modification of flight plans to meet the needs of any specific flight operation.

*Typical Systems*

*Performance Data Computer System*
This system provides advisory data in alphanumeric format on a

CRT display, in addition to the positioning of target command 'bugs' on a Mach/airspeed indicator and EPR indicators, such indicators operating on electrical servomechanism principles. Provision is also made for interfacing the system with autothrottle and automatic flight control systems. A schematic diagram of the system, which consists primarily of a control and display unit, computer and mode annunciator, is shown in Fig. 8.17.

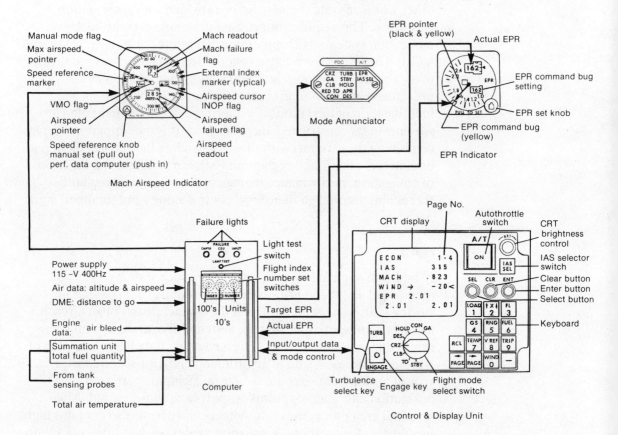

**Fig 8.17** Performance Data Computer System

Abbreviations are extensively used for the display of data by the control and display units of this and other flight management computer systems, and these abbreviations/acronyms and their definitions are given in Appendix 9.

*Control and Display Unit (CDU)*
The CDU provides the major input link to the system and allows the flight crew to make inputs to obtain EPR and airspeed displays and can also be used for obtaining decision-making data in relation to an aircraft's flight profile. The CRT has a 2 in × 3 in screen and enables data to be displayed over a 13 (column) × 6 (row) matrix.

The selection of EPR and airspeed data for various phases of flight is accomplished by a flight mode select switch, the modes and associated displays being as follows:

| | |
|---|---|
| TO | Take-off EPR limits for the outside air temperature entered by the flight crew |
| CLB | EPR and speeds for the desired climb profile; best economy, maximum climb rate, or crew-selected speeds |
| CRZ | EPR and speeds for the desired cruise schedule; best economy, long-range cruise, or crew-selected speeds |
| DES | Descent speed, time and distance for best economy |
| HOLD | EPR, speed and endurance for holding |
| CON | Maximum continuous EPR limits for existing altitude, temperature and speed |
| GA | Go-around EPR limit for existing altitude, temperature and speed |

The standby (STBY) position of the select switch is used for data entry and for an automatic check-out of the system.

The function of the 'ENGAGE' key is to couple the target command 'bugs' of the Mach/airspeed indicator and EPR indicators to computer command signals which drive the bugs to indicate the speed and EPR values corresponding to those displayed on the CRT screen. If the data is verified by the computer to be valid, engageable and different from the data presently engaged, the engage key illuminates and is extinguished after engagement takes place; at the same time the appropriate light of the mode annunciator is illuminated.

The key marked 'TURB' is for use only in cruise and when turbulent flight conditions are to be encountered. When pressed it causes the CRT to display the appropriate turbulence penetration data, i.e., airspeed in knots (also Mach number at high altitudes), pitch attitude and the $N_1$ percentage rpm. In the turbulence mode, the target command speed and EPR 'bugs' engage automatically. This mode is disengaged by pressing the key a second time or else engaging another flight mode.

In order that the flight crew may load keyboard-selected data into the system, three push-button switches are provided above the keyboard for SELecting, CLeaRing and ENTERing data. In connection with the selection and entering of data, question marks and two symbols are displayed at the right-hand end of a data line; a caret (<) and an asterisk (*). The caret signifies that the computer is ready to accept data, while the asterisk signifies that the data next to it may be entered or changed if necessary.

The keyboard primarily serves a dual function in that it

(1) permits the flight crew to enter pure numeric data into the computer and (2) permits desired performance function data to be called up from the computer for display. The data appropriate to the keys is given in Table 8.3 and is displayed in the form of pages, each page being numbered in the top right-hand corner. For example, the page shown on the CDU in Fig. 8.17 is page 1 of a set of four relating to 'economy fuel' in the cruise mode. In order to call up each of the remaining pages the $\overrightarrow{\text{PAGE}}$ key is successively pressed. Similarly, the $\overleftarrow{\text{PAGE}}$ key permits cycling of the pages in reverse order. When a flight mode or performance function is first selected, the first page of a set is always automatically displayed.

The RCL key is used whenever a performance function is being displayed and if it is required to recall a display corresponding to a selected flight mode.

The two switches in the upper right-hand corner of the CDU are associated with auto-throttle system operation. When the A/T Annunciator switch is pressed, an internal light is illuminated to indicate connection of the auto-throttle system and at the same time an 'EPR' light in the mode annunciator is illuminated. The PDCS then adjusts the throttles to track the EPR target values displayed on the CDU and by the command bugs of the associated EPR indicators. In order for the auto-throttle system to adjust engine power in relation to indicated airspeed, the second switch 'IAS SEL Annunciator' is operated; the system then drives the throttles so as to track the speed target values displayed on the CDU and by the command bug of the Mach/airspeed indicator.

*Computer*

The computer is of the hybrid type, and the inputs, outputs and unit interfaces are as shown in Fig. 8.17. Program storage is by means of a PROM and an additional non-volatile memory for retaining all entered data during any interruption of the power supply. Built-in test equipment circuits and software operate continuously to check all critical circuits of the system. The fuel summation unit which is a component of the PDCS, develops an a.c. voltage signal that is proportional to the total fuel on board the aircraft; the signal being a combination of those produced by the fuel-quantity-indicating system sensing probes which are located in the fuel tanks.

Failure lights on the front of the computer indicate whether a fault is in the computer, CDU or input signals. The INDEX NUMBER switches, which are of the rotary type, are used for

**Table 8.3** Performance function data

| Key | Data pages displayed |
|---|---|
| LOAD 1 | Outside air temperature, destination airport elevation, reserve and alternate fuel and zero fuel weight for the intended flight plan |
| ↓ × ↑  2 | Flight level intercept data for use in solving time and distance problems in climb and descent |
| FL 3 | Economy cruise and long range cruise speeds at appropriate flight level |
| GS 4 | Present ground speed computed from a known true airspeed (TAS) and wind component |
| RNG 5 | Total flight endurance as well as distance/time solutions to fuel reserve at any flight level |
| FUEL 6 | Fuel for engaged cruise speed (economy, long range and manually entered speeds). Used only in the CRZ and CON modes |
| TEMP 7 | Ambient temps (TAT & SAT) TAS, and temperature deviation from ISA |
| V REF 8 | Reference landing speeds for various flap settings, based on aircraft's correct gross weight |
| TRIP 9 | Optimum initial cruise flight level for inserted trip distances |
| WIND 0 | Data in respect of automatically computed and manually inserted wind components |
| — | Negative value data and also 'test pages' while in 'STBY' mode |

Microelectronics in Aircraft Systems

programming a flight index number from 0 to 200 into the computer so that maximum economy flight modes are modified according to time-related costs compared to fuel costs. The switches are guarded to eliminate the possibility of inadvertent changing of the index number.

*Mode Annunciator*

This unit indicates the flight mode driving the command bugs of the EPR and Mach/airspeed indicators. The legend appropriate to a flight mode is illuminated when the mode is selected, and the 'ENGAGE' or 'TURB' button switches are depressed.

*Typical Display*

The number of data pages associated with all the performance functions and flight modes it is possible to select on the CDU is quite considerable, and limitations on space do not permit a full description of each to be given here. However, the fundamentals of presentation and data entry methods may be understood from the example given in Fig. 8.18

In this case, the first page of the fuel performance function for the maximum economy speed schedule is displayed, and the question marks on the second line (Fig. 8.18(a)) indicate that the computer is asking for the distance to go from the present position of the

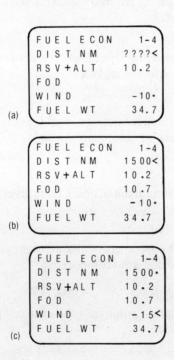

**Fig 8.18** Example of PDCS Display

aircraft. The caret indicates the computer's readiness to accept this information. The CLR button switch is then pressed and this causes erasure of the question marks and the caret to flash on and off to indicate that the numbers appropriate to the distance should be 'keyed in' to the display from the keyboard—1500 nautical miles in this case (Fig. 8.18(b)). When this has been done the ENTER button switch is then pressed, following which the computer goes through an input validity check routine. If the input is valid, the caret will stop flashing, thereby advising that the data has been accepted. An INVALID message is displayed if the input exceeds any limitation. The computer also computes the fuel over destination (FOD) value and causes it to be displayed. If the computer requires more data on another line the caret drops to that line automatically. The asterisk against the wind component of −10 knots signifies that any change to its value may be effected. If, for example, the wind speed has increased to 15 knots (the minus sign indicates a headwind) then it may be entered by first pressing the SEL key which causes the caret and the asterisk to exchange positions (Fig. 8.18(c)). The CLR key is then pressed, the new wind speed value is keyed in and finally displayed by pressing the ENTER button switch.

*Flight Management System*

A flight management system (FMS) is currently the most advanced of systems, providing as it does full integration of all the functions referred to earlier (see page 172) and which are necessary to fly optimized flight profiles either in manual or fully automatic control modes. The system is a union of autonomous and generally asynchronous units interconnected by a network of ARINC data busses to satisfy specific functional needs. In many cases redundant units are present to meet requirements for functional availability, flight safety or aircraft coverage.

Figure 8.19 illustrates schematically the computing units which are typically a formal part of an FMS and also how by means of the data busses they communicate with the principal elements of the system, namely the flight management computer (FMC) and its associated control and display unit (CDU). By virtue of their communication link these two units are together designated as a flight management computer system and this provides the primary interface between the flight crew and the aircraft. Inputs from other interfacing systems and sensors are also transmitted to the data busses, but for reasons of clarity have been omitted from Fig. 8.19.

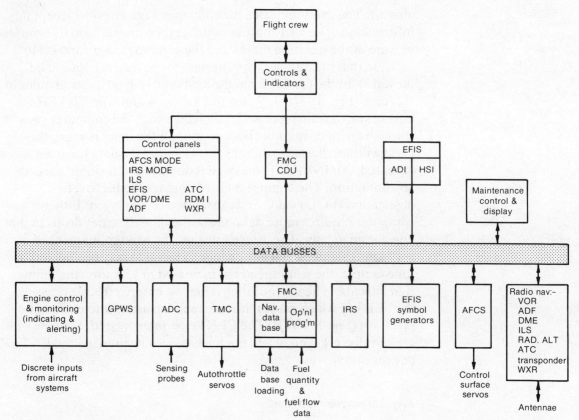

**Fig 8.19** Flight Management System

*Flight Management Computer*

Typically, a computer incorporates three different types of memory: a bubble memory for holding the bulk navigation and aircraft performance characteristics data bank; a C-MOS RAM for holding specific navigation and performance data, and the active and secondary flight plan, all 'down-loaded' from the bubble memory; and a UV-PROM for the operation program, which may be reprogrammed at card level.

The data base which is used for all computations contains numerous types of records in memory and these are given in Table 8.4. All the data is unique to each aircraft operator, depending on the routes flown, and is initially programmed on magnetic tape. The tape cartridge is inserted into a portable data base loader unit which, after connection to the computer, is operated so as to transfer the data to the bubble memory. Any subsequent changes in navigation aids and procedures, and route structure changes, are also incorporated in the data base by means

**Table 8.4** Records in data base memory

| Record | How identified and defined |
|---|---|
| Radio-nav aids: VOR, DME, VORTAC, ILS, TACAN | Identifier ICAO region, latitude and longitude, frequency, magnetic declination, class (VOR, DME, etc.), company defined figure of merit,* elevation for DME, ILS category, localizer bearing |
| Waypoints | Each waypoint defined by its ICAO region, identifier, type (en-route, terminal), latitude and longitude |
| En-route airways | Identified by route identifier, sequence number, outbound magnetic course |
| Airports | Each identified by ICAO four-letter code, latitude and longitude, elevation, alternate airports |
| Runways | Each identified by ICAO identifier, number, length, heading, threshold latitude and longitude, final approach fix identifier, threshold displacement |
| Airport procedures | Each identified by its ICAO code, type (SID, STAR, profile descent, ILS, RNAV), runway number/transition, path and termination code |
| Company routes | Origin airport, destination airport, route number, via code (SID, airway, direct, STAR, profile descent, approach), via identifier (SID, name, airway identifier, etc.), cruise altitude, cost index |

* The figure of merit is a number assigned to each navigational aid to indicate the maximum distance at which it can be tuned.

of the data loader, in accordance with a specified time schedule, e.g., a 28-day cycle.

*Control and Display Unit*

The CDU of one example of an FMCS which is currently in use is shown in Fig. 8.20. It is basically similar to that adopted for the performance data computing system described earlier, but in keeping with the role to be played by an FMCS the unit is much more sophisticated in respect of data selection and corresponding displays.

The operation of the keyboard function keys is summarized in Table 8.5. As in the case of a PDCS display unit, annunciators are also provided but form an integral part of the unit.

While on the ground, the flight crew can construct a detailed flight plan by inserting data in selected data pages and, in conjunction with the comprehensive data base stored in the computer, the plan is raised to active control status. In flight, the system receives data from the relevant aircraft sensors and radio-navigational aids, and then as the flight proceeds it presents the flight plan in a progressive

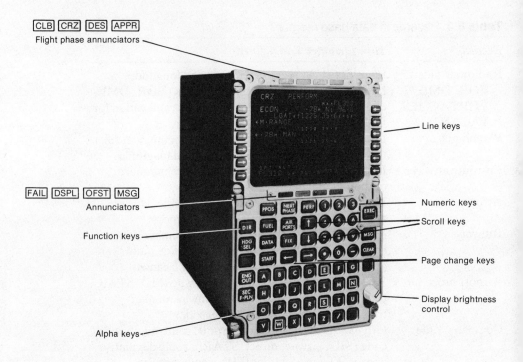

**Fig 8.20** FMCS Control and Display Unit

**Table 8.5** Summary of operation of keyboard function keys

| Key | Selection and data displayed |
|---|---|
| P POS | Returns display to show the active navigation leg page, i.e. the aircraft's present position |
| NEXT PHASE | Changes a navigation leg display to the beginning of the next phase of the flight plan |
| PERF | Selection of performance pages |
| DIR | Permits direct entry of revisions to flight plan from present position to any waypoint |

**Table 8.5** – contd

| Key | Selection and data displayed |
|---|---|
| FUEL | Selection of fuel pages |
| AIR-PORTS | Displays the navigation legs page which includes the next airport along the current flight plan |
| HDG SEL | Selection of headings to be flown automatically via the FMS |
| DATA | Displays data index pages relevant to: lateral, vertical, performance, key waypoints, sensors, maintenance, navigation, aircraft configuration, history |
| FIX | To check or up-date aircraft's position |
| START | Selects 'START DATA REQUIRED' pages for flight crew to initiate and construct flight plans |
| ENG OUT | Presents performance data pages relating to engine out operation |
| SEC F-PLAN | Selects secondary flight plan facility for re-clearance or return-leg planning |
| EXEC | To promote a temporary plan to active status. The bar illuminates when the FMCS has enough data to create an active plan, and remains illuminated until the temporary plan has been executed, or cancelled by pressing P POS |
| MSG | Informs computer that any message displayed on CDU has been acknowledged and the message will either be stored or erased |
| CLEAR | To delete incorrect scratch pad entries |

'scroll' form. Pitch, roll and thrust demands are also computed and in communicating these to an AFCS and a TMS accurate control of the flight profile and maximum fuel economy can be provided. Various pages of data can be selected for review by the flight crew at any stage of a flight, and predictions concerning its future phases can be assessed by inserting detailed revisions, the future implications of which are computed and displayed. In addition to its own display unit, the FMCS has also the unique capability of presenting navigation data in 'changing map' form on the display unit of an electronic flight instrument system (see also Chapter 6).

The flat-faced CRT of the display unit gives a dual character size presentation with 24 stroke-written characters per 14 lines. Small-size characters signify data with default values, or computer predicted values which can be changed by the flight crew when the data is being supplied by the computer; large-size characters signify data entered by the flight crew.

*Data Pages and Flight Plan Construction*

Many pages of data can be accessed and displayed, and space does not permit them all to be shown here. However, some indication of character presentation and method of entering data in general may be understood by considering the example of the display shown in Fig. 8.21, which is used to initiate the construction of a flight plan when the 'START' key is pressed.

The data lines adjacent to the line keys constitute the 'operational area' of the display and can be accessed by line key selection for entry or revision of flight plan information or for selection of

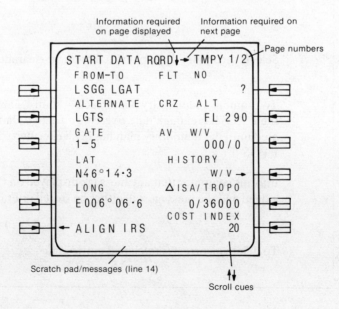

**Fig 8.21** Example of a page display

184 Computers

displayed options. The first 12 character spaces on the bottom line (line 14) are used as a scratch pad for information entered by the flight crew via the alphanumeric keys, the next 10 spaces are reserved for FMS messages to the crew and the last two spaces are reserved for scroll cues signified by upward- and downward-pointing arrows. When the appropriate scroll key on the keyboard is pressed the display moves up or down for the purpose of reviewing the display.

Arrows which appear against characters of a display indicate flight crew options, which may be the choice of display data or may result in some functional activity of the system such as aligning the inertial reference system (IRS) as indicated in Fig. 8.21. The choice of option is signified by pressing the line key adjacent to an arrow; this results in a change to the content of an existing page and erasure of the arrow.

Requests for data entry are indicated by question marks. For example, in Fig. 8.21 the request is for the flight number, and in response this is first entered into the scratch pad by using the alphanumeric keys and then pressing the line key adjacent to the question mark. If the format of the entry is correct, the flight number appears in large-size characters on the appropriate line and the scratch pad clears. If the format is incorrect, the word ERROR appears in the message space, the MSG annunciator illuminates and the incorrect number remains in the scratch pad. After an incorrect entry has been attempted the ERROR legend and MSG annunciation are acknowledged by pressing the MSG key. The CLEAR key is then pressed to delete the entire scratch pad entry.

It is possible to change any data by over-writing with new values from a scratch pad entry. If, for example, it is required to change the displayed data base cruise altitude, the new value is 'keyed' into the scratch pad and then transferred to the altitude line by pressing the adjacent line key. Similarly, if the airport terminal gate from which an aircraft is to depart is changed, the new gate number may be over-written in the display. The FMC will automatically enter the revised latitude and longitude values for the new gate to which the IRS must be re-aligned.

The TMPY legend on the title line of start data pages signifies that the data being entered by the flight crew relates to the construction of a temporary flight plan.

Flight plan construction and associated changes of data pages are essentially in two sections. Firstly, the navigation section involves insertion and acceptance of route number, airport codes, cruise altitude, latitude and longitude data and IRS alignment. When this has been accepted by the computer the bar in the EXEC key is illuminated and this section of the plan is executed by pressing the key. A page is then displayed requesting data needed to construct the second section of the plan which relates to fuel on board, reserve fuel, trip fuel and time, weights and centre of gravity position. When this has been computed, a 'START COMPLETE' page is displayed to show the relevant fuel, weight data and time components, and the EXEC key is again pressed so that the whole of the flight plan is raised to 'active' status.

*Performance and In-flight Displays*

The next step is to initiate the display of data pages relevant to the performance section of the flight plan, and this is done by pressing the PERF key. These pages are concerned with management of engine thrust, pitch attitude and other alternative modes of performance control limited to the various phases of flight. The normal mode is 'economy' by which the most economical climb, cruise and descent speeds are computed. Each flight plan has its own performance scroll of pages related to those of the navigation legs scroll. The first page is the take-off performance page in which the flight crew must enter the values of the take-off speeds $V_1$ and $V_R$.

After take-off and during climb at the appropriate climb speed, the CLB annunciator is illuminated and the take-off performance page is automatically replaced by a climb performance page. At the top of the climb, the CRZ annunciator is illuminated and a cruise performance page is then presented to display continuously up-dated information related to optimum performance and destination arrival fuel and time. During cruise descent a descent performance page is displayed, but the point at which it is presented depends on the point at which the descent is commenced. For example, if the descent is commenced prior to the planned point in the vertical profile (by authorizing it via the AFCS control panel) the cruise performance page is first replaced by a cruise descent page and the aircraft descends at the selected vertical speed. When the aircraft captures the planned descent profile the display then changes to the descent performance page and the DES annunciator is illuminated. If the descent is commenced at the planned profile point, the cruise performance page is replaced directly by the descent performance page and the CR DES annunciator is illuminated. During approach, and when the leading edge slats are extended, an approach performance page replaces the descent performance page and the APPR annunciator is then illuminated.

If any revisions are made while in flight, the TMPY legend will also appear on the appropriate data page being displayed to indicate that a complete temporary flight plan is generated. The aircraft continues under the control of the active flight plan until the temporary plan is raised to active status by pressing the EXEC key. If it is to be aborted, the P POS key is pressed.

*Data Index*

An index of data is contained on a page which can be called up for display by pressing the data key. When the line keys adjacent to the titles are pressed, further pages are presented to display the information noted in Fig. 8.22.

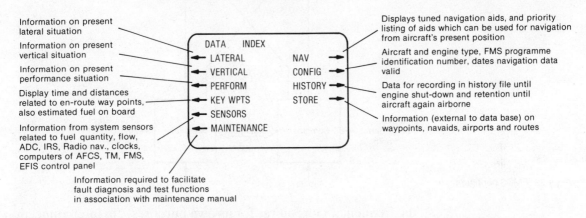

**Fig 8.22** Data index page

*System Configuration*

Two FMC systems are installed in an aircraft, each having its own CDU situated on the centre console (see Figs 6.1 and 6.2) and each controlling its associated automatic flight control system, auto-throttle system and radio-navigational aids. The basic configuration is shown in Fig. 8.23. In normal operating conditions, both computers operate together and share and compare each other's information, i.e., they 'cross talk' by means of an interconnecting data bus. The pilots can operate their displays independently for review or revision purposes without disturbance to the active flight plan and without affecting the other CDU commands. Typically, a working arrangement would be for the performance pages to be displayed on one pilot's CDU, while the other pilot's CDU would display navigation legs.

When a temporary flight plan is created in one system the other system (referred to as the 'offside' system) has no access to this plan, but it can review the active plan in the normal way. In addition, the offside system is inhibited from creating a temporary flight plan until the previous temporary state is cancelled or raised to active.

Each computer has its own VOR/DME receiver and determines

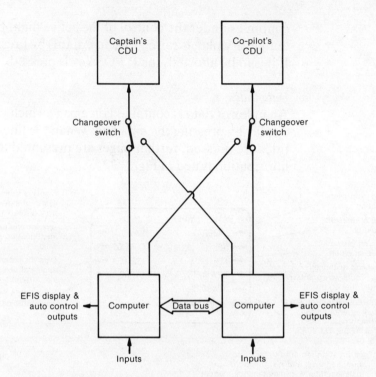

**Fig 8.23** FMCS configuration

the frequencies it requires for its own purposes. No interconnection between systems is possible except when a lateral revision is effected at the present position of the aircraft.

In the event of failure of one computer, each pilot has the means whereby he can select his own CDU into the other system.

# 9 Aircraft systems

The systems and instruments described in this chapter are some further examples of those utilizing digital circuit techniques.

**Altitude Reporting System**

The control of air traffic along the many air corridors in the vicinity of major airports is dependent on rigid procedures for communication between individual aircraft and ground control stations in order that traffic may be identified and assigned to requisite separation levels. In addition to normal voice transmissions, the communication procedure involves the use of an airborne transponder which, in response to interrogation signals from a radar transmitter/receiver at the air traffic control centre, automatically transmits coded reply signals to the centre. The signals are then computer-processed, decoded, and then alphanumerically displayed to the air traffic controller on his primary radar screen.

Aircraft altitude is one of the important parameters required to be known, and to further reduce time-consuming voice transmissions, a method of automatically transmitting such data from an altimeter was devised and also became a mandatory feature of the air-to-ground communication procedure.

The interrogation system as a whole forms what is termed Air Traffic Control Secondary Surveillance Radar, and can operate in four modes of interrogation: A, B, C and D. Modes A and B are used for identification, Mode C for altitude reporting, while Mode D is at present unassigned. In each case the interrogating signal, which is transmitted on a frequency of 1030 MHz from a rotating directional antenna, comprises a pair of pulses $P_1$ and $P_3$, and in order that the airborne transponder can 'recognize' in which mode it is being interrogated, the pulses are spaced at different time intervals. The spacing is taken from the leading edge of the first pulse to the leading edge of the second (Fig. 9.1). It will be noted from diagram (a) that a third pulse $P_2$ can also be transmitted from a control antenna; its purpose is to suppress side lobe radiation from

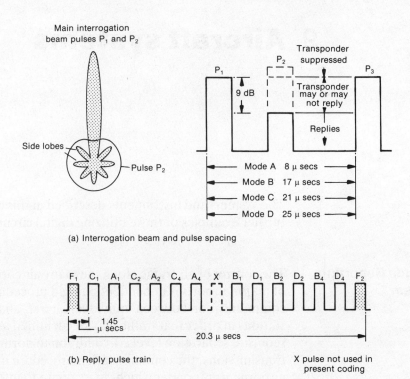

**Fig 9.1** Pulse structure and transponder operation

the interrogator antenna, and to ensure that the transponder replies only to the main beam directional signal pulses. This is effected by a 'gating' circuit which compares the relative amplitudes of the pulses and enables the transponder to determine whether the interrogation is a correct one, or due to a side lobe.

When the transponder decodes the interrogating signal it will reply by transmitting a train of information pulses on a frequency of 1090 MHz, and in a coded sequence which depends not only on the interrogation mode, but also on pre-allocated code numbers which, for operation in Modes A and B, are selected by the pilot on the transponder control unit. In Mode C operation, code numbers are automatically transmitted by the transponder, which also receives signals corresponding to specific altitudes from an altimeter and in a manner to be described later.

The train of information pulses from the transponder may consist of up to 12 pulses spaced 1·45 μs apart, depending on the reply code selected on the control unit (Fig. 9.1(b)). The pulses lie between two additional framing pulses $F_1$ and $F_2$, which are fixed at a spacing of 20·3 μs, and which are always transmitted in the reply. In a 12-pulse train the number of codes available is $2^{12} = 4096$, the codes being numbered 0000 to 7777, the latter giving all 12 pulses when four selector knobs of the control unit (Fig. 9.2(a)) are

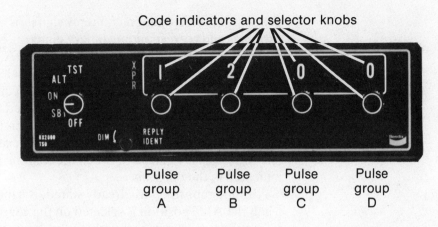

| Code number selected | A B C D | A B C D | A B C D | A B C D | A B C D |
|---|---|---|---|---|---|
|  | 0 0 0 0 | 0 0 3 2 | 3 4 0 0 | 0 3 6 7 | 7 7 7 7 |
| Reply pulses | Framing pulses only | $C_1$ $D_2$ $C_2$ | $A_1$ $B_4$ $A_2$ | $B_1$ $C_2$ $D_1$ $B_2$ $C_4$ $D_2$ $D_4$ | $A_1$ $B_1$ $C_1$ $D_1$ $A_2$ $B_2$ $C_2$ $D_2$ $A_4$ $B_4$ $C_4$ $D_4$ |

| Code digit | Pulse assignment | | | |
|---|---|---|---|---|
|  | A | B | C | D |
| 0 | None | None | None | None |
| 1 | $A_1$ | $B_1$ | $C_1$ | $D_1$ |
| 2 | $A_2$ | $B_2$ | $C_2$ | $D_2$ |
| 3 | $A_1$ $A_2$ | $B_1$ $B_2$ | $C_1$ $C_2$ | $D_1$ $D_2$ |
| 4 | $A_4$ | $B_4$ | $C_4$ | $D_4$ |
| 5 | $A_1$ $A_4$ | $B_1$ $B_4$ | $C_1$ $C_4$ | $D_1$ $D_4$ |
| 6 | $A_2$ $A_4$ | $B_2$ $B_4$ | $C_2$ $C_4$ | $D_2$ $D_4$ |
| 7 | $A_1$ $A_2$ $A_4$ | $B_1$ $B_2$ $B_4$ | $C_1$ $C_2$ $C_4$ | $D_1$ $D_2$ $D_4$ |

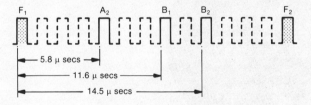

**Fig 9.2** Reply pulse coding

correspondingly set. Each control knob controls a group of three pulses as shown; the letters in this case designate pulse groups and not interrogation modes. The first control knob controls the A group, the second knob the B group, and so on. The subscripts to each letter of a group are significant since their sum equals the digit selected on the control unit, and this may be seen from the basic code table and examples given in Fig. 9.2(b). Since the whole system is based on a digital computing process, the encoding and decoding of interrogation and reply signals is dependent on logic variables and corresponding binary digits or 'bits'. In the example shown in Fig. 9.2(c) the code 2300 in mode A has been selected on the control unit and this produces the equivalent bit groups 010, 110, 000 and 000 respectively. The selection of the digit 2 causes only pulse 2 of the A group to be transmitted, while the selection of digit 3 causes the transmission of pulse 1 plus pulse 2 of the B group. Thus, the reply pulse train in this example would consist of only three pulses spaced between the framing pulses at the intervals indicated in the diagram.

Altitude reporting, as already stated, is a mode C operation, and when the ALT position is selected on the control unit of the transponder the latter will reply to the corresponding interrogation signal as well as to a mode A interrogation signal. However, whereas in mode A operation the pulse trains of reply signals are associated with manually selected codes, in the altitude reporting mode pulse trains are produced automatically and supplied to the transponder by an encoding altimeter or, in some cases, by the altitude measuring section of an air data computer. The arrangement of an encoding altimeter is shown in Fig. 9.3 and from this it will be noted that an encoder assembly is mechanically actuated by the aneroid capsule assembly in addition to the conventional pointer and mechanical digital counter display mechanism.

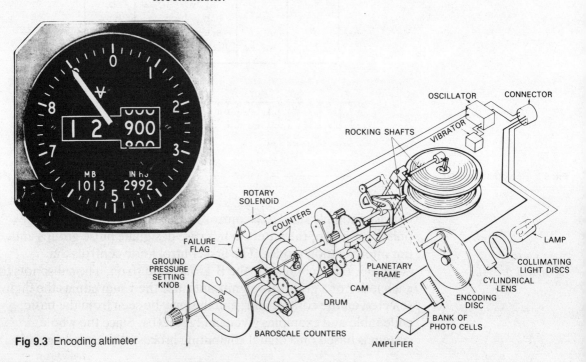

**Fig 9.3** Encoding altimeter

The encoder assembly is of the optical type consisting of a light source, light-collimating discs, a cylindrical focusing lens, encoder disc, a bank of photoelectric cells, and an amplifier. The encoder disc (Fig. 9.4) is made of glass and is etched with transmitting and

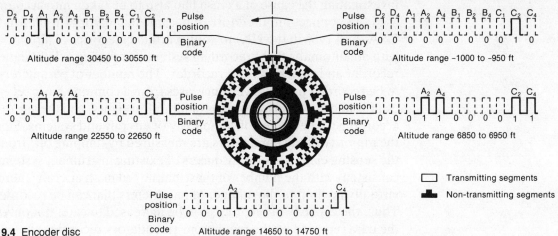

**Fig 9.4** Encoder disc

non-transmitting segments arranged in 11 concentric rings and spaced so as to produce Gray-coded pulses corresponding to 100 ft (30·48 m) increments of altitude. The coding is in accordance with that established by ICAO and set out in Annex 10— Aeronautical Telecommunications. The disc makes a single turn to produce the requisite number of counts compatible with the altitude range for which the instrument is designed.

During operation, light from the 14 V lamp passes through the collimating discs which produce parallel rays that are then focused through the cylindrical lens onto the encoding disc in the form of a sharp line of light. Depending on the aircraft's altitude at any one time, the disc and segments along a particular radial will be at some corresponding position with respect to the bank of photoelectric cells. The cells will accordingly respond to the position of the transmitting and non-transmitting segments and produce outputs which are then amplified and supplied to the transponder. Examples of the pulse codings produced are shown in Fig. 9.4.

**Flight Data Recording**

The recording of certain of the parameters measured by instruments, and of any unusual incidents connected with engines, airframes and performance generally, has always been a feature of in-flight management procedures. The earliest method of recording, and one which is still mandatory, is that requiring entries to be made in the aircraft's 'tech log'. However, as aircraft and their systems became more sophisticated in their operating procedures, the number of parameters to be monitored increased, and this imposed limitations on the 'tech log' entry method. Furthermore, the question of how to retrieve data which would be of value in investigating the cause of a crash had also to be taken into account.

In this connection, therefore, it became mandatory (in 1958 in the USA, and 1965 in the UK) to equip certain categories of aircraft with an automatic flight recording system comprising a flight data recorder and a cockpit voice recorder. The number of parameters to be recorded progressively increases from a common basic set established according to groups of aircraft weights, and therefore according to the degree of complexity of aircraft and systems. Since the majority of the parameters are measured by 'tapping off' from the sensing elements or transducers of existing instrument systems, consistent with the number of these employed in an aircraft, there is virtually no limit to the number of parameters that can be recorded. Thus, the capacity of recorders can be increased to cater not only for the data required by the regulations (mandatory recording) but also for the acquisition of data which can be of value to an aircraft operator in planning for economic operation, and for the monitoring of aircraft, engines and systems reliability (non-mandatory recording).

Since the mandatory requirements for aircraft to be equipped with flight data recorders relate primarily to the acquisition of data which will prove valuable during investigations into the cause of a crash, the overall development of data recording systems has been built up using such data acquisition as the foundation.

The relative value of flight recording has, from investigators' experience, been indicated in respect of two very general categories of crash. The first category is that in which some form of aircraft, or system, malfunction has been the primary cause. In attempting to establish the primary failure, analysis of the wreckage would probably play the principal role, and this would be supported by such 'pre-crash' recorded data as time, altitude, speed and acceleration. The second category is that in which the crash has an operational cause, i.e., one in which no engineering defect or deterioration in performance of the aircraft has occurred. Some examples of this are loss of control resulting from improper handling by the flight crew, adverse weather conditions, or navigational factors. In such a crash, the role of wreckage analysis would be limited to establishing data concerning the aircraft's final configuration, and broad details of impact attitude and speed. The role of the flight data recorder in this case would be more prominent, since it would provide a time history of the principal manoeuvres throughout the flight.

*Parameters Selected*

In the selection of the mandatory parameters for crash investigation purposes, the objective is to obtain, either directly or by deduction from the recorded data, the following information:

1   the aircraft's flight path and attitude in achieving that path;
2   the basic forces acting on the aircraft, e.g., lift, drag, thrust and control forces;
3   the general origin of the basic forces and influencing factors, e.g., status of such systems as primary and secondary flight controls, hydraulic power supply, cabin pressurization, electrical power, and navigation.

The mandatory parameters are specified in national regulations for civil aircraft operation and generally relate to the following:

(a)   time (GMT or elapsed);
(b)   indicated altitude;
(c)   indicated airspeed;
(d)   vertical (normal) acceleration;
(e)   magnetic heading;
(f)   pitch attitude;
(g)   roll attitude;
(h)   flap position;
(i)   engine power.

**Electromagnetic Recording**

In this method, the analog signal data from the transducers appropriate to each parameter are first encoded, i.e., transformed into a digital coded form in accordance with the binary scale of notation. The encoded signals then pass through such sub-channels as conditioning, logic, multiplexing, and modulation, and are fed as a series of binary coded pulse currents to an electromagnetic recording or 'writing' head. The magnetic core of the 'writing' head has a very small air gap (typically 25 $\mu$m or 0·001 in) between pole pieces, and is located in very close proximity to the recording medium which, in some recorders, may be a plastic-based tape coated with ferrite material, while in others, it may be a stainless steel wire. The air gap is kept as small as possible to permit high packing density, and furthermore, in conjunction with the small

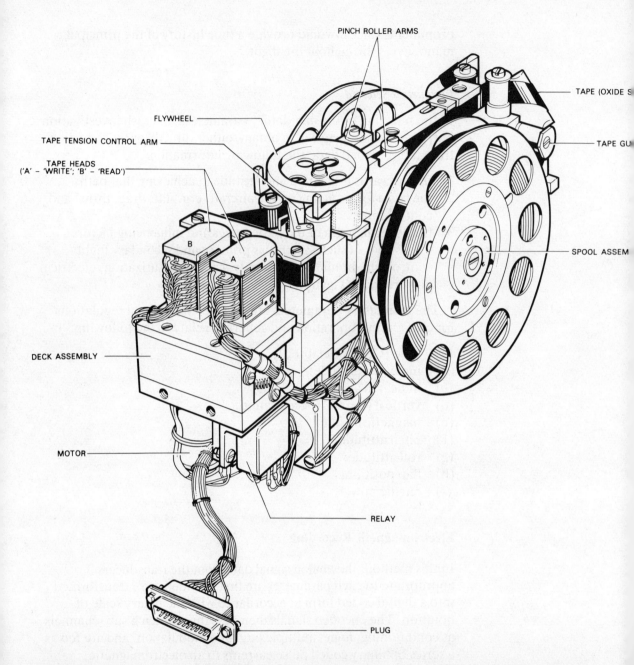

**Fig 9.5** Example of transporting magnetic tape

space between the gap and recording medium surface, it minimizes any interference the generated flux might have on adjacent sections of the medium already magnetically coded.

The tape or wire is wound on spools and guide rollers, and is transported over the 'writing' head by a synchronous motor supplied with 115 V a.c. The transport rate is typically 44·4 mm (1·75 in) per second for magnetic tape and 63·5 mm (2·5 in) per second for wire.

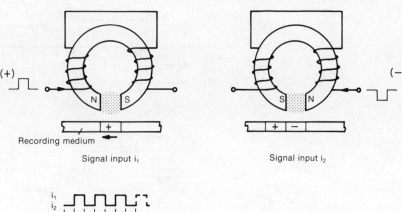
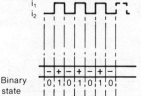

**Fig 9.6** 'Writing' head operation

An example of one method of transporting magnetic tape is illustrated in Fig. 9.5.

Whenever a current pulse is applied to the coil of the 'writing' head, a leakage flux fringes out between the pole pieces of the core, and as the tape or wire is transported past the pole pieces, magnetically polarized 'spots' or sections are produced in a serial manner on the tape or wire, the polarities being in accordance with the binary coded pulse currents applied to the coil. Thus, and as shown in Fig. 9.6, binary words are polarized on the tape or wire, to provide a 'magnetic store' of information. The number of 'bits' which can be stored in a unit length of tape or wire gives the packing density; some typical values are 900 bits per inch for tape, and 414 bits per inch for wire. In magnetic tape recording, coded data are stored on a number of tracks into which a tape is divided; in the example shown in Fig. 9.6, the tape is 12·7 mm (½ in) wide and 260 m (850 ft) long. The 'writing' head has a corresponding number of magnetizing coils, each of which is automatically selected to provide sequential track recording.

The technique of electromagnetic recording is further illustrated in Fig. 9.7. To provide correct interpretation of the stored information when 'reading' the tape or wire, the signal inputs to the magnetizing coils of the 'writing' head must first be assembled into an identifiable format. This is done by multiplexing the pulses in a signal conditioning unit, so that the format comprises a frame of 64 words, each word containing 16 bits; a frame is of 1 s duration. The first word in each frame is a frame synchronizing (or relative time)

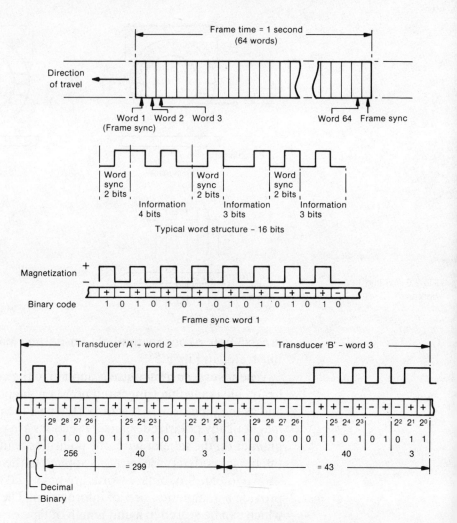

**Fig 9.7** Electromagnetic recording

signal comprising a repetitive series of 1 s and 0 s, the remaining 63 words being available for pure information on the required parameters. Each of these words includes three interposed two-bit (01) word synchronizing signals which divide the pure information into groups of 4 bits, 3 bits and 3 bits respectively. Thus, each word contains 10 information bits, and as per binary notation, each word has an equivalent full-scale decimal count of 1024 ($2^{10}$). In the example in Fig. 9.7, the ten information bits of word 2 from a transducer A have been magnetically encoded 0100, 101 and 011; conversion to decimal requires only the 1 s to be raised to higher powers of 2 and always from right to left, so that the code therefore converts to 256 + 40 + 3 and so is equal to 299. Similarly, the word 3 from transducer B converts to 43 in decimal notation.

Information on all parameters is sampled within the signal conditioning unit of the flight recorder system at periodic intervals to construct the multiplex pattern, some parameters being sampled more frequently than others. Some details of sampling rate and word allocation based on one type of wire recorder in current use are given in Table 9.1.

**Table 9.1** Sampling rates of recorded parameters

| Parameter | Sampling rate (per second) | Word number allocation |
|---|---|---|
| Time | 1 | 1 |
| Acceleration | 8 | 6, 14, 22, 30, 38, 46, 54, 62 |
| Pitch | 8 | 2, 10, 18, 26, 34, 42, 50, 58 |
| Altitude | 1 | 9 |
| Airspeed | 1 | 13 |
| Heading | 1 | 29 |
| Flight number and date in three groups of 10; 1 at → and 2 at → | 2 <br> 1 | 3, 5, 35, 37 |

For identification purposes, it is necessary to record flight reference information, i.e., flight number and date. This is done by operating selector switches on a flight encoder panel located in the flight crew compartment. The switches are numbered, and when selected they feed an output to the signal conditioning unit for multiplexing and transmission to the recorder in groups of binary coded information.

**Flight Data Acquisition Systems**

Except for the requirements laid down for flight testing and certification procedures, the acquisition of a wide range of data additional to that associated with accident investigation is non-mandatory. It can, however, provide an aircraft operator with the means of monitoring the performance of systems on a comparative basis, and from any variations to obtain advance warnings of possible failures of a component or a system overall. For example, from the monitoring of parameters related to engine performance, i.e., pressures, temperatures, vibration levels, an engine 'signature' can be detected over a period of time and this can be used as a performance datum against which the recorded parameters of other engines can be compared each time a readout of acquired data is made. Flight data acquisition systems are therefore installed in various types of large public transport aircraft, and operate in conjunction with a flight recorder. Figure 9.8 illustrates an example

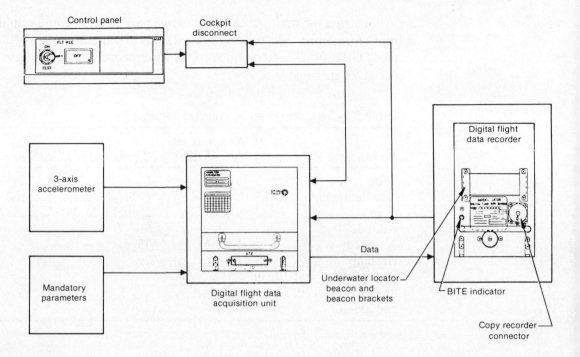

**Fig 9.8** Digital flight data acquisition and recording system

of one such system (based on that adopted in the Boeing 757), and the parameters which are monitored are shown in Fig. 9.9.

## VOR System

A VOR (v.h.f. omnidirectional radio range) system is a short-range navigational aid comprising ground-based transmitting stations, or beacons, and airborne receiving elements. It provides en-route information on the bearing of an aircraft from the points at which the stations are geographically located. The stations are spaced at intervals of 50 to 80 nautical miles within what is termed the 'airways system'.

A VOR station transmits a very high frequency (v.h.f.) carrier wave operating in the 108–118 MHz band, and on which are superimposed two low-frequency modulating signals. One of these signals, known as the reference signal, is of fixed phase, and is radiated in all directions (hence the term omnidirectional), while the other is a rotating beam signal and varies in phase to produce an infinite number of 'radials'. Thus, at any particular point relative to a station (which is lined up on magnetic north) a specific phase relationship exists between the two signals, and this is indicated in Fig. 9.10. Each station is identified by a coded signal which is received by the navigation receiver in the aircraft when the corresponding station transmitting frequency has been selected. The display of bearing information is presented on an indicator mounted

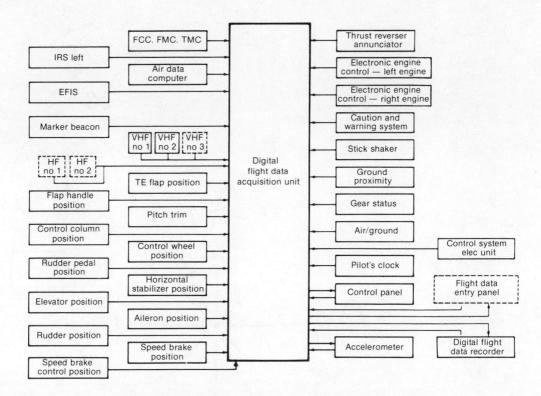

**Fig 9.9** Parameters monitored by flight data acquisition system

on the aircraft's main instrument panel. The type of indicator depends on the equipment specified for a particular type of aircraft; for example, it may be an omni-bearing indicator, a radio magnetic indicator, or it may be a horizontal situation display indicator which forms part of a flight director system. In general, however, the display has three main indicating elements: (i) a bearing scale or an index, which is positioned by a selector knob to indicate the radial on which the aircraft is to be flown; (ii) a pointer indicating whether the aircraft is flying on the radial or has deviated from it; and (iii) a TO–FROM indicator to indicate whether the aircraft is flying towards a station or away from it. A warning flag is also provided to indicate 'system off', receiving of low-strength signals, or no signals at all.

Figure 9.10 illustrates an omni-bearing indicator, and this serves as a useful example in understanding the fundamentals of VOR operation as a navigational aid. After a station has been identified from its coded signal and tuned-in, the omni-bearing selector knob is rotated to set the required bearing on the bearing scale, and also to position the rotor of a resolver synchro within the indicator. The navigation receiver which continuously compares the phases of the transmitted signals, and compares them in turn with the phase shift produced in the resolver synchro, supplies an output signal to a

Microelectronics in Aircraft Systems 201

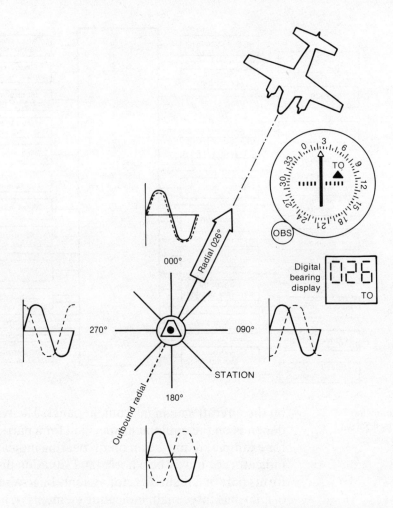

**Fig 9.10** VOR system
——— Reference phase;
– – – Variable phase

meter movement controlling the vertical pointer of the indicator. When the pointer lies in the centre of the indicator it indicates that the phasing of the required bearing radial signals has been matched and that further rotation of the selector knob is unnecessary. The reading shown on the bearing scale at that instant indicates the heading on to which the aircraft must be turned in order to fly along the corresponding radial to the station.

At the same time, the TO–FROM indicator indicates TO usually by means of an arrow-shaped pointer. With any departure from the radial as a result of, say, crosswind drift, the deviation pointer will be deflected to one or other side, and so command a turn in the direction of the deflection in order to intercept the radial once again. As the aircraft approaches overhead the station, so the deviation pointer meter movement becomes more sensitive as a result of the convergence of the radials, and its indications, together with those of the TO–FROM indicator, become erratic. In this part of the approach, overhead the station and in departing from it, the aircraft is said to be in the 'cone of confusion'. The extent of the cone varies with altitude and groundspeed, typically from a few seconds at low altitude, to as much as two minutes at high altitude. Outbound flight from the station is indicated by the TO–FROM indicator changing to FROM and, provided an accurate heading is maintained, the deviation pointer continues to give corrective command information for the interception of the outbound radial.

In addition to the selection of omni-bearings in the manner just described, automatic selection and display of such data are also effected by means of digital circuit techniques, and so the associated microelectronic devices are applied to the majority of current radio navigation systems. Figure 9.11 illustrates the logic circuit of an example based on one of the many systems available, and shows a typical application of such devices as counters, shift registers, and a microprocessor.

The purpose of the phase-locked loop and divide-by-3600 counter (diagram (a)) is to generate a system clock pulse that has a resolution of $0.1°$. The reference frequency for the loop is the 30 Hz VOR station reference phase, and the frequency of the loop itself is 108 kHz. This is determined from the sine wave of the reference phase (Fig. 9.12). Since this wave has a period of 33·3 ms, the period per degree is $33·3/360 = 0·0925$ ms. Converting this to microseconds and tenths of a degree, we obtain a period of 9·25 ms per $0·1°$. The reciprocal of time gives us the loop frequency which is equal to 1000/9·25 or 108 kHz. This is supplied as the clock input to the counter, and also as an input to the AND gate.

The counter is a 12-bit ripple type and provides a 30 Hz phase comparison signal input to the phase-locked loop. This signal is derived by the counter dividing the loop frequency of 108 kHz by 3600, i.e., the pulses needed to cater for the bearings from 0 to 360°. The outputs $Q_5$, $Q_{10}$, $Q_{11}$ and $Q_{12}$ are clamped by diodes to 8 V through a pull-up resistor. The reset line is also clamped to 8 V. When the reset line is logic 1, the counter is in the reset state. At the beginning of the count, the reset line is clamped to logic 0 by the $Q$ outputs which are also logic 0 at this time. As the count progresses, and until it reaches 3600, there is always at least one $Q$ output at logic 0 and so the reset line always remains at logic 0. At pulse number 3600, all the $Q$ outputs of the counter go to logic 1, the clamping diodes are now reverse-biased, and so the reset line is pulled up to 8 V, thereby resetting the $Q$ outputs to logic 0. The diodes again clamp the reset line to logic 0 and the count is then recycled.

As indicated earlier, the bearing to a VOR station is the phase

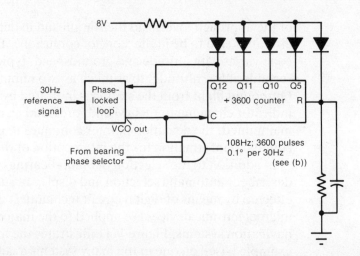

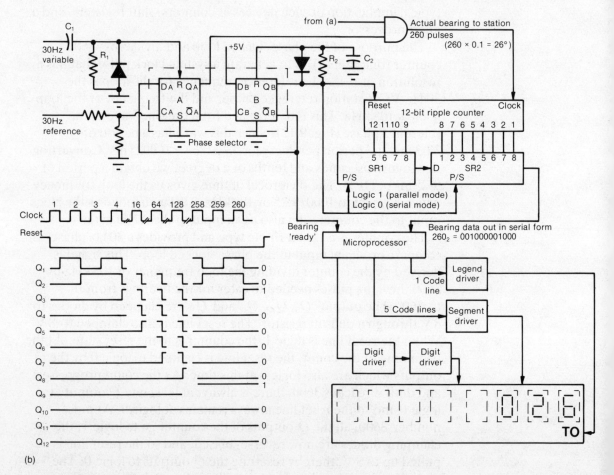

**Fig 9.11** Automatic selection and display of VOR data

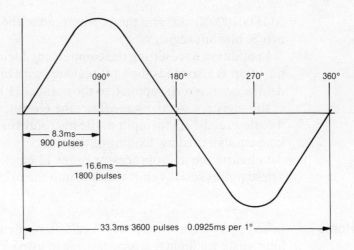

**Fig 9.12** Sine wave of VOR reference phase

difference between the reference and the variable phase 30 Hz signals. Let us assume (as shown in Fig. 9.10) that an aircraft is flying to a station on a bearing of 026°. If we refer once again to Fig. 9.11(b), it will be noted that the reference phase signal is supplied as a clock pulse to flip–flop A of the bearing phase selector, the purpose of which is to pass on the number of 0·1° pulses equivalent to the phase difference. The variable phase signal is differentiated by capacitor $C_1$ and resistor $R_1$ and is supplied as a reset signal to the flip–flop.

The resetting of flip–flop A establishes a logic 0 and a logic 1 at $Q_A$ and $\overline{Q}_A$ respectively, the logic 1 being supplied as the clock pulse for flip–flop B. As a result of this pulse the output $Q_B$ will also be logic 1 and is looped back to the reset of flip–flop B, and is delayed for 30 s by capacitor $C_2$ and resistor $R_2$. The reason for this delay is to allow $Q_B$ to remain at logic 1 long enough to act as a reset pulse for the 12-bit counter. $\overline{Q}_B$ is connected to the $D_A$ input of flip–flop A and goes to logic 0 on the reset pulse to A; 30 s later when A resets, $\overline{Q}_B$ goes to logic 1. Thus flip–flop A is now 'primed', and on the leading edge of the clock pulse, $Q_A$ goes to logic 1 and $\overline{Q}_A$ goes to logic 0. The output from $Q_A$ which detects the correct phase angle and represents the bearing to the VOR station is applied as an input to the AND gate. The other input to the gate is from the phase-locked loop as described earlier. Therefore, for a 26° phase angle, 260 pulses will be gated through and applied to the clock input of the 12-bit ripple counter. As the count progresses, the $Q$ outputs change state according to the timing diagram and stop at pulse 260, as shown in Fig. 9.11. (At this stage, compare the waveforms $Q_1$ to $Q_4$ with those of the 4-bit counter shown in Fig. 5.19.) The outputs $Q_1$ to $Q_{12}$ yield the number 260 in binary code 001000001000 and this is presented to the parallel inputs of the two 8-bit shift registers.

In addition to resetting the counter, the logic 1 output $Q_B$ of flip–flop B also loads the shift registers with binary coded bearing data, since it is also supplied to the parallel (P) and serial (S) inputs of the registers. At the same time, the $\overline{Q}_B$ output is supplied as a 'bearing ready' signal input to the microprocessor, and is delayed long enough to allow loading of the registers with the coded bearing data before the microprocessor issues 12 clock pulses to them. These pulses serially shift the data into the microprocessor.

**Multiplex Systems**

In large passenger transport aircraft, in particular those operating long-distance flights, it is customary to provide passengers with entertainment services such as multi-channel selection of stereo or monaural music, and 'in-flight movies'. The services are primarily under the control of the cabin attendants from their control panels, and once the appropriate reproducing systems are in operation, music and movie audio is 'piped' to service units at every passenger seat location. Each unit contains a socket for a stethoscope type headset and multi-channel selector switch together with switches for reading lights and 'attendant call'. An override system is also provided to enable a cabin attendant or pilot to interrupt the audio programme so that public-address (PA) announcements will also be heard by passengers using headsets. If conventional electrical methods were to be used for the interconnection of all essential elements of a system, a considerable quantity of cabling and associated connectors would be required. Audio signals are therefore transmitted in digital format and in a time-division multiplexing mode (see page 71) via a data highway consisting of a single coaxial cable. A simplified block diagram of the system is given in Fig. 9.13.

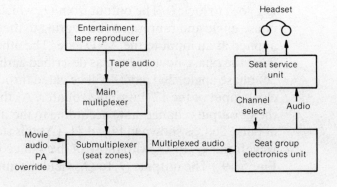

**Fig 9.13** Application of multiplexing

The channel arrangement in this example is one in which the tape reproducer has 10 audio channels, one from each of four movie projectors in different zones of the passenger cabin, and one channel from the PA system. The main multiplexer samples the audio signals from the tape reproducer during a channel time, and by the method of pulse-amplitude modulation reconstructs the signal. This signal is then converted into a 12-bit pulse-coded signal, the bits being combined with clock bits and stereo tag bits. The multiplexer also supplies the timing for channels 1–4 and 15, and timing and data for channels 5–14. Pulse-code modulation is used to combine the 10 audio channels for transmission over a single cable, each channel being sequentially switched into a sampling circuit. A 12-bit coded sample represents the pulse amplitude of the sample taken. The sample from each channel is placed sequentially on the output data line together with clock pulses and stereo or monaural identification to direct the channels to the correct earplugs of the headset. A frame synchronizing pulse follows channel 15 and precedes channel 1 time.

The output from the main multiplexer is connected by coaxial cable to sub-multiplexers in the appropriate zones of the passenger cabin. These multiplexers contain circuits necessary (i) to convert audio-frequency analog signals from the movie projectors or PA amplifier, into an equivalent pulse-coded modulation signal, (ii) to multiplex the four channels of zone-generated monaural audio with the 10 channels of music, and (iii) to maintain system synchronization. A 12-bit binary code is generated from each audio channel by an A/D converter, and this places the bits on the data line with the data from the sub-multiplexers.

Data outputs from the sub-multiplexers are routed in a series/parallel manner through seat group electronics units which consist of demultiplexer and seat decoder circuits. The demultiplexer extracts the data from the digital pulse train based on the programmes selected by passengers at the seat control units, and by means of a D/A converter the coded signals are converted to audio. The decoder circuits decode stereo channel pairs of audio signals for the left and right earplugs of a headset. The circuits also decode any PA announcement to inhibit all programme selection for the duration of the announcement.

**Servo-operated Instruments**

In a number of aircraft the instruments designed for the measurement of such parameters as engine speed and exhaust gas temperature are of the servo-operated type. Since the measurements are in analog form, for purposes of signal processing and amplification their circuits incorporate operational amplifiers performing some of the functions described in Chapter 4.

The instruments consist essentially of a d.c. servomotor which, in response to the difference between demand signals generated by speed or temperature sensors as appropriate, and by a position feedback potentiometer, drives a pointer and a 3-digit counter. The signals are processed through solid-state circuits forming signal processing, servo-amplifier and power supply card modules (see Fig. 1.19).

As an example of operational amplifier application let us consider the servo-amplifier module circuit shown in basic form in Fig. 9.14.

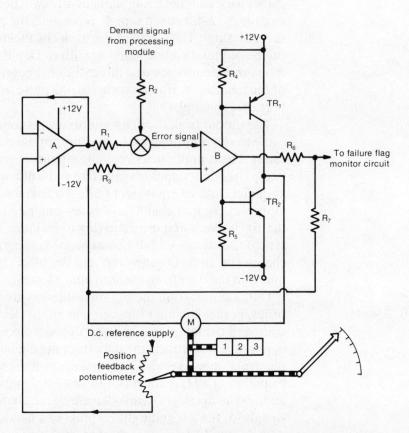

**Fig 9.14** Application of operational amplifiers

The module consists of two operational amplifiers: A which is connected as a voltage follower, and B which operates in conjunction with two power transistors $TR_1$ and $TR_2$ to serve as the actual servomotor drive amplifier. After processing by the signal processing circuit module, a demand signal is applied to a summing point between the output of amplifier A which is a position feedback signal, and the inverting input of amplifier B. If there is any difference between the feedback and the demand signals, the input to B is an error signal.

The quiescent state current drawn by amplifier B from the 12 V d.c. supply passes through resistors $R_4$ and $R_5$, but under zero error signal conditions it is not large enough to forward-bias either of the power transistors. Assume that a change in the measured parameter occurs, e.g., there is an increase in engine speed. Then the signal processing circuit will produce an 'increase' demand signal and apply it to the summing point, which will result in a negative polarity (with respect to 0 V) error signal being applied to the inverting input of amplifier B. The current flowing through $R_4$ will therefore be increased owing to the additional current drawn by the amplifier through its load resistors $R_6$ and $R_7$. The voltage drop across $R_4$ will now be large enough to switch on $TR_1$, which will drive the servomotor, pointer and counter toward an increased speed value represented by the demand signal. At the same time, the wiper of the feedback potentiometer will also be driven, and through the non-inverting input of amplifier A and its feedback loop, an output is produced to 'back off' the demand signal until the error signal is zero. At this stage, the instrument pointer and counter will display the demanded speed.

When a demand signal corresponding to a 'decrease' in engine speed is applied to the summing point, an error signal of opposite polarity to that noted above will be applied to the inverting input of amplifier B. Transistor $TR_2$ will now be switched on through the voltage drop across $R_5$, and so the servomotor will be driven until the error signal is again zero.

Resistor $R_1$ provides a load for operational amplifier A, while $R_3$ minimizes any effects of offset voltage and thermal 'drift' of operational amplifier B.

Figure 9.15 is a schematic diagram of a servo-operated instrument designed for the indication of engine rpm but, unlike the instrument just described, it utilizes a microprocessor for driving the pointer and also an LED display of the dot-matrix type (see also page 88). The percentage rpm input signals, which are generated at a frequency comparable to speed by the engine-driven sensor, are fed to a multiplexer together with a 'bite' oscillator output and, under the control of the microprocessor, the appropriate input is selected for measurement by a frequency measuring logic circuit module. In order to permit more rapid attainment of high resolution of frequency measurement, the logic circuit module contains a multiplier consisting of a phase-locked loop which increases the input frequency from 70 Hz to approximately 10 kHz at 100% rpm, a cycle counter and a period counter. The output from the multiplier

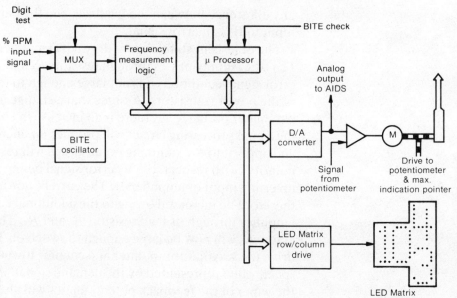

**Fig 9.15** Servo-operated instrument with LED matrix display counter

is fed to the cycle counter, which starts to count from a zero transition of the input frequency. At the same time, the period counter, which has a fast clock (2 mHz), starts counting, and when it reaches the desired count for the resolution required it carries on counting until the next zero transition of the input frequency, at which point both counters are stopped. The cycle counter then has the number of cycles, while the period counter has the number of fast clocks over which the measurement took place. This method of measurement provides essentially a measurement time that is independent of signal input frequency.

The output thus derived by the frequency measuring logic module is used by the microprocessor to calculate a digital output equivalent to the percentage rpm to be displayed by the servo-driven pointer, and also the matrix format for the LED display row/column drive. The pointer is driven by a d.c. servomotor and so, as will be noted from Fig. 9.15, the microprocessor output must be supplied to the motor amplifier via a D/A converter.

Each numeral of the dot matrix display is generated within a $9 \times 5$ matrix, and there is one blank row and two blank columns between numerals, the latter spacing allowing for the insertion of a decimal point when required. The microprocessor is configured such that as the input signal changes with engine rpm it 'addresses' the matrix drive, which illuminates the LEDs in the rows and columns so that they form the numerals corresponding to the instantaneously measured values; the numerals, in turn, appear to roll smoothly

**Fig 9.16** 'Rolling digit' sequence (courtesy Smiths Industries Ltd)

upwards or downwards in the same way that a mechanical drum counter changes its indications. The three displays shown in Fig. 9.16 are based on a three-digit matrix configuration, and are given as an example of the 'rolling digit' sequence in an increasing rpm situation.

In order to ensure legibility of the LED display under varying ambient light conditions, the intensity of the light emitted by the diodes may be adjusted. A method most commonly adopted for this purpose is one in which the matrix drive is integrated with a centralized system under the control of light-sensitive devices (photo-diodes) located on the flight-deck main instrument panel.

Other facilities provided for by the instrument are:

1. a digits test whereby the complete matrix of the LED display can be energized to check that all the diodes illuminate;
2. the supply of an analog output of the percentage rpm signal to an airborne integrated data system (AIDS); and
3. a test switch for checking the operational status of the motor-driven pointer together with the LED display, and of the AIDS analog output

# 10 Handling of microelectronic circuit devices

The voltage and current requirements of microelectronic devices are of a very low order of magnitude. It is therefore necessary to observe strict precautions to avoid damage or destruction when carrying out functional testing and trouble-shooting procedures. Apart from the effects of improper testing procedures, however, there are some devices whose circuits can, by the very nature of their construction, be damaged or destroyed by static electricity discharges resulting simply from the manner in which the devices are physically handled. These devices are referred to as *electrostatic-sensitive devices* (ESD); the relative sensitivities of those most specifically concerned are given in Table 10.1.

**Table 10.1** Sensitivities to static electricity of some common devices

| Device | *Electrostatic discharge range where damage can occur (V)* |
|---|---|
| Field effect transistors (MOSFET) | 100–200 |
| JFET | 140–10 000 |
| Complementary metal oxide silicon (CMOS) | 250–2000 |
| Schottky diodes, TTL | 300–2500 |
| Bipolar transistors | 380–7000 |
| Precision thin-film resistors | 150–1000 |
| Emitter coupled logic (ECL) | 500 |
| Silicon-controlled rectifiers (SCR) | 680–1000 |

**Static Electricity**

Static electricity is, as the name implies, electricity at rest, and is a phenomenon the effects of which are well known. Most people have experienced an 'electric shock' when, say, opening a room door. This effect is due to the fact that, under certain conditions, the body becomes positively or negatively charged; e.g., as a result of just walking across the room. When the hand is in close proximity to the door handle an opposite polarity charge is induced in the handle. If, at that moment, the potential difference is great enough it will cause electrons to jump the gap from the negative charge to the positive charge, and thereby produce a momentary surge of current across the gap. It is this current which produces the 'electric shock' sensation. The effect is therefore analogous to that produced by a capacitor; the body and door handle form the plates, and the air in the gap the dielectric.

How does this relate to ESDs? Let us assume that we pick up a printed circuit board assembly of such devices and make contact with the edge connector contacts or some other exposed part of the circuit. If there is a great enough potential difference between the hand and the assembly there will be a surge of current to produce the electric shock sensation, and the assembly will have become oppositely charged. When the assembly is being inserted in its appropriate 'black box', there will again be a surge of current, but this time from the assembly to the black box and without any shock sensation being experienced. On carrying out a functional check, however, it will be found that the complete unit is either 'down' on its performance or not functioning at all. Removal of the assumed 'faulty' board assembly for subsequent testing reveals that one, or maybe several, of the microcircuit packs has failed. Any effects from the electrostatic discharge would not, of course, be visible to the naked eye, but examination of the circuit under an electron scanning microscope at the appropriate magnification would indicate rupturing of the oxide film, in some cases resembling the 'bomb crater' effect shown in Fig. 10.1.

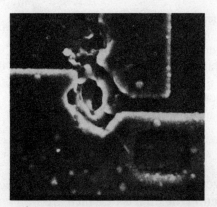

**Fig 10.1** Effect of electrostatic discharge

Table 10.2 gives some idea of the extent to which static charges can be generated by the human body. The values are based on data obtained under some typical industrial conditions.

**Table 10.2** Typical static voltages

| Situation | Relative humidity of air | |
| --- | --- | --- |
| | Low 10–20% (V) | High 65–90% (V) |
| Walking across a carpet | 35 000 | 1500 |
| Walking over vinyl floor covering | 12 000 | 250 |
| Worker at bench | 6 000 | 100 |
| Vinyl envelopes containing work instructions | 7 000 | 600 |
| Polythene bag picked up from bench | 20 000 | 1200 |
| Work chair padded with urethane foam | 18 000 | 1500 |

Charges can, of course, be generated on a variety of materials and parts (we say 'on' because static electricity is a surface phenomenon) simply by bringing surfaces into contact and sliding them against each other. Electrons are shared between the surfaces and when they are pulled or moved apart; the electrons do not have an opportunity to return to their original positions, thereby creating varying degrees of charge of differing polarities dependent on the type of materials in contact. Static electricity can always bleed away of its own accord; but in the real world of manufacturing microelectronic devices, handling them during assembly and testing, replacing board modules of a unit installed in an aircraft, etc., the time available for carrying out the processes involved is very much shorter than the static bleed-off time. In consequence, therefore, static is an ever-present hazard and so its effects must be guarded against at all times.

The first level of protection for devices is normally provided by buffer and/or bypass circuitry designed into the devices, but such circuitry can be limited in its application so that there can be no guarantee of its providing complete and permanent protection. Some measure of protection can also be afforded once devices are installed in their respective printed circuit boards and when these in turn are positioned in their appropriate 'black boxes'. These, however, may have to be removed and replaced during routine inspection and trouble-shooting procedures and, as already noted, any accidental contact with exposed parts of a circuit can lead to static charging and discharging and resultant failures. The only remaining practical recourse, therefore, is to observe strictly all the precautions and procedures laid down in all documentation relating not only to the handling of individual devices, but also to the equipment assemblies in which they are used. These precautions and procedures apply at all stages from the source of manufacture through storage, transportation, testing, installation and removal while in service.

**Identification**

An obviously important requirement is the identification of the packaging containing ESDs and of any assembly, be it a card module or complete electronic 'box', which may utilize ESDs in the circuitry. For this purpose, therefore, special decals are affixed to packaging and assemblies; some examples of decals currently in use are shown in Fig. 10.2. In cases where the connector pins of an electronic 'box' may be susceptible to a discharge, an additional decal is affixed near the connector as a warning to personnel not to touch the connector pins.

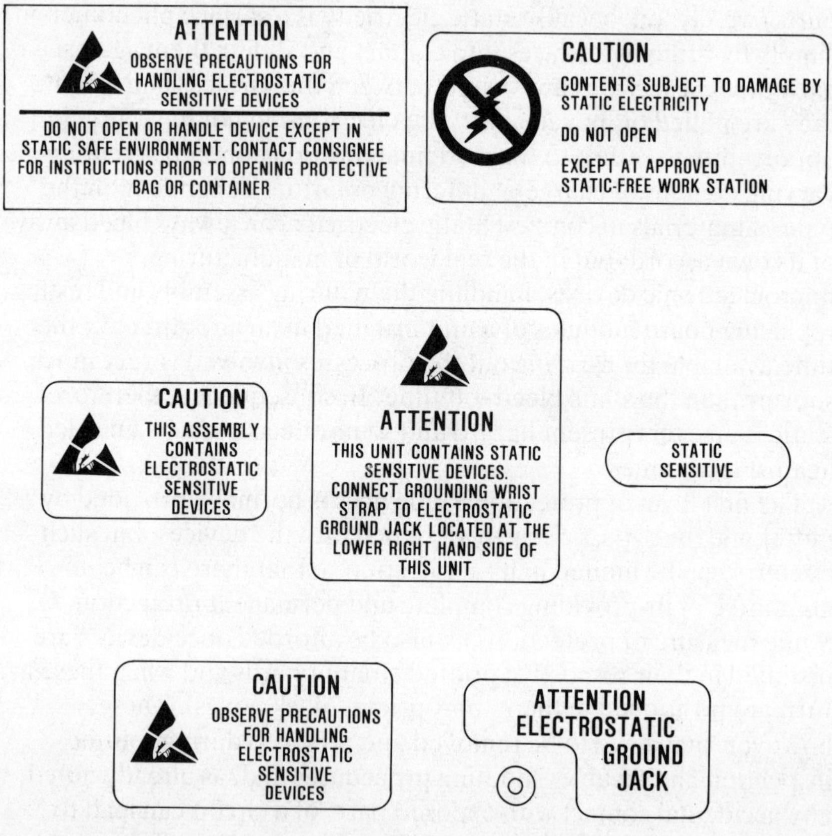

```
ATTENTION
OBSERVE PRECAUTIONS FOR
HANDLING ELECTROSTATIC
SENSITIVE DEVICES
DO NOT OPEN OR HANDLE DEVICE EXCEPT IN
STATIC SAFE ENVIRONMENT. CONTACT CONSIGNEE
FOR INSTRUCTIONS PRIOR TO OPENING PROTECTIVE
BAG OR CONTAINER
```

```
CAUTION
CONTENTS SUBJECT TO DAMAGE BY
STATIC ELECTRICITY
DO NOT OPEN
EXCEPT AT APPROVED
STATIC-FREE WORK STATION
```

```
CAUTION
THIS ASSEMBLY
CONTAINS
ELECTROSTATIC
SENSITIVE
DEVICES
```

```
ATTENTION
THIS UNIT CONTAINS STATIC
SENSITIVE DEVICES.
CONNECT GROUNDING WRIST
STRAP TO ELECTROSTATIC
GROUND JACK LOCATED AT THE
LOWER RIGHT HAND SIDE OF
THIS UNIT
```

```
STATIC
SENSITIVE
```

```
CAUTION
OBSERVE PRECAUTIONS
FOR HANDLING
ELECTROSTATIC
SENSITIVE
DEVICES
```

```
ATTENTION
ELECTROSTATIC
GROUND
JACK
```

**Fig 10.2** Identification decals

**Protection and Packaging**

Wherever there are charge-sensitive devices there will also be the problem of protecting them during transportation and storage, and so specialized packaging is essential for individual devices, card modules and complete electronic 'box' assemblies. The packaging for devices and card modules takes the form of bags made from a material which is quasi-conductive, i.e., a material whose surface or volume resistivities are too high to be conductive, but conductive enough to 'bleed off' charges in no more than a few milliseconds. Various types of material are used in the manufacture of bags and some typical examples are:

1. polyolefin film impregnated with carbon;
2. polyethylene resin impregnated with an organic anti-static substance which migrates to all surfaces rendering them quasi-conductive; the additive allows the material to remain transparent and from the colour of the material is usually referred to as 'pink poly';
3. aluminium layer sandwiched by an outer layer of anti-static-treated spin-bonded polyolefin and an inner layer of 'pink poly'; and
4. cushion-pack bags made from 'pink poly' foam material.

Other protective measures involve the shorting of connecting leads or pins of devices by means of wire, spring clips, metal foil or by inserting the leads or pins into a conductive foam material which must always be of a type that will not induce corrosion of leads or pins. For PCBs and card modules having edge connectors, specially formed strips called 'shunts' are placed over connectors to keep them all at the same potential and also to protect them against physical damage.

For complete electronic 'box' assemblies, covers or caps made from a conductive material are placed over the connectors. In cases where covers of some other non-conductive material are used, the covers are treated by spraying on a liquid conducting film. Since this has a certain 'life', the date of treatment should always be indicated on the protective covers.

## Static-free Work Stations

It is of little use to provide static-free packing if the protected contents are not removed in a static-free environment; thus, purpose-designed work stations within special handling areas must always be used as instructed by the identification decals. Typically, a station consists of a conducting work surface connected to either a conducting mat or floor surface which, in turn, is connected to ground. The operative's chair or stool is also provided with a conductive seat cover and grounded. A grounded strap which an operative must place around his or her bare wrist is also provided and has to be worn when handling devices and assemblies (see Fig. 10.3). In order to neutralize any random charges on devices or

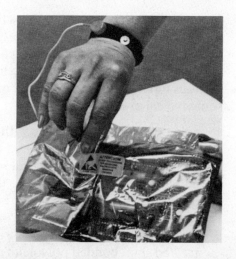

**Fig 10.3** Use of grounding wrist strap

assemblies during those periods when other protection is removed, the work station is also equipped with an ionizer which generates positive and negative ions and blows air so-charged across the work area and operative's hands. All test data sheets, specifications and other such documentation required during testing, assembly, etc., has to be contained within anti-static envelopes.

## Handling Outside Special Handling Areas

For the engineer responsible for the maintenance of avionic systems and equipment installed in an aircraft, handling will obviously be done outside any special handling areas, and then it will only be related to the removal and installation of the specified line replaceable units (LRUs) such as card modules or complete electronic boxes containing ESDs. It is nonetheless essential that precautions appropriate to in-service maintenance tasks be observed.

Before removing a card, electrical power sources should be isolated and a wrist strap must be worn by the engineer; the strap is connected to a convenient grounding point on the unit containing the card or, in aircraft using 'card file' units, the strap is connected to an electrostatic jack point provided on the units (see Fig. 10.3). The card should then be removed by means of its extractors and immediately inserted in a conductive bag which should be sealed as appropriate to the particular design of bag, and identified by a decal. Staples or adhesive tape should not be used for sealing a bag. To maintain integrity of the bag during subsequent transportation, it should be placed in a rigid container which should also be identified by a decal.

The precautions regarding power isolation, use of wrist strap and card extractors, apply equally to the installation of a replacement card which should always remain in its conductive bag until ready for actual installation.

When removing or installing complete electronic boxes or other metal-encased, rack-mounted components, electrical power sources must also be isolated, and great care must be taken when removing and replacing anti-static covers or caps from connectors not to touch the pins. Unless otherwise specified, it is not necessary to wear a wrist strap, since the metallic mass of the component can serve as a ground for discharging any static accumulated by the person carrying out the removal or installation procedure.

# Appendix 1

## Abbreviations associated with microelectronic components, circuits and systems

In the documentation dealing with the characteristics of components, principles of circuit operation and applications, etc., extensive use is made of abbreviations as a form of shorthand communication. Their meanings are not always defined in the text of a document, and so reading one, particularly for the first time, can prove a little disconcerting. Space does not permit the inclusion of every abbreviation in current use, but the reader should find the following of help.

| | |
|---|---|
| AIM | Avalanche-induced migration |
| Al-gate MOS | Aluminium-gate metal-oxide semiconductor |
| ALRS | Arithmetic-logic register stack |
| As | Arsenic |
| ASCR | Asymmetrical silicon controlled rectifier |
| BBD | Bucket brigade device |
| BCSL | Base-current switch logic |
| BeAMOS | Beam-addressed metal-oxide semiconductor |
| BFET | Bipolar field-effect transistor |
| BFL | Buffered field-effect transistor logic |
| BIGFET | Bipolar insulated-gate field-effect transistor |
| BIMOS | Bipolar metal-oxide semiconductor |
| BML | Bipolar memory, linear |
| BOP | Bipolar operational power |
| BORAM | Block-oriented random-access memory |
| BSR | Bit shift register |
| BU | Buffer unit |
| CAM | Content-addressed memory |
| CASH | Charge-amplified sample and hold circuit |
| CATT | Controlled-avalanche transit time |
| CCCL | Complementary constant-current logic |
| CCD | Charge-coupled device |
| CCL | Collector-coupled logic; also constant-current logic |
| CDI | Collector diffusion isolation |
| CERDIP | Ceramic dual-in-line package |
| CERMET | Ceramic metallic |

| | |
|---|---|
| CMI | Compound monolithic integration |
| CML | Current-mode logic |
| CMOS | Complementary metal oxide semiconductor (or C-MOS) |
| CMRR | Common-mode rejection ratio |
| COS/MOS | Complementary-oxide silicon/metal-oxide silicon |
| CROM | Control and read-only memory |
| CSL | Current-sinking logic |
| CTD | Charge-transfer device |
| CTL | Common (also complementary) transistor logic |
| | |
| DAC | Digital-to-analog converter |
| DAR | Data-access register |
| DC | Discrete component |
| DCC | Discrete component circuit |
| DCFL | Direct-coupled field-effect transistor logic |
| DCL | Direct-coupled logic |
| DCTL | Direct-coupled transistor logic |
| D/G | Driver/gate |
| DIC | Digital integrated circuit |
| DIL | Dual-in-line |
| DIP | Dual-in-line package |
| DMA | Direct memory access |
| DMC | Direct memory channel |
| DMOS | Double-diffused metal-oxide semiconductor |
| DSC | Digital-to-synchro converter |
| DSM | Dynamic scattering mode |
| DTL | Diode–transistor logic |
| DUF | Diffusion under epitaxial film |
| | |
| EAROM | Electrically alterable read-only memory |
| EBCDIC | Extended binary-coded decimal interchange code |
| ECL | Emitter-coupled logic |
| ECTL | Emitter-coupled transistor logic |
| EE-PROM | Electrically erasable programmable read-only memory |
| EFL | Emitter follower logic |
| EPROM | Erasable programmable read-only memory |
| | |
| FAMOS | Floating-gate avalanche-injection metal-oxide semiconductor |
| FED | Field-effect device |
| FET | Field-effect transistor |
| FIFO | First-in first-out |
| FP | Flat pack |
| FPLA | Field-program logic array |
| F-PROM | Field-programmable read-only memory |
| | |
| Ga | Gallium |
| GaAsFET | Gallium–arsenic field-effect transistor |
| Ge | Germanium |

| | |
|---|---|
| GFET | Gate field-effect transistor |
| GTOSCR | Gate and turn-off silicon controlled rectifier |
| HCMOS | High-density complementary metal-oxide semiconductor |
| HIC | Hybrid integrated circuit |
| HIIC | High-isolation integrated circuit |
| HLTTL | High-level transistor–transistor logic |
| HMOS | High-performance metal-oxide semiconductor |
| HNIL | High-noise immunity logic (HiNil) |
| HTFD | Hybrid thick-film device |
| HTL | High-threshold logic |
| IC | Integrated circuit |
| IG-FET | Insulated gate-field-effect transistor |
| IHS | Integrated heat sink |
| IIL | Integrated injection logic ($I^2L$) |
| ILP | Integral lead package |
| IMPATT | Impact avalanche transit time |
| I/O | Input/output |
| J-FET | Junction field-effect transistor |
| LAPUT | Light-activated programmable unijunction transistor |
| LARAM | Line-addressed random-access memory |
| LASCR | Light-activated silicon controlled rectifier |
| LIC | Linear integrated circuit |
| LIFO | Last-in first-out |
| LL | Logic level |
| LOCMOS | Local oxidation complementary metal-oxide semiconductor |
| LSI | Large-scale integration |
| LSIA | Large-scale integration array |
| MAR | Memory-address register |
| MCA | Multichip array |
| MEAL | Micro-extended assembly language |
| MIC | Monolithic integrated circuit |
| MICROM | Micro instruction control read-only memory |
| MIR | Memory information register |
| MOSFET | Metal-oxide silicon field-effect transistor |
| MOSIC | Metal-oxide silicon integrated circuit |
| MOST | Metal-oxide silicon transistor |
| MRV | Maximum reverse voltage |
| MSI | Medium-scale integration |
| MST | Monolithic system technology |
| MTL | Merger transistor logic |
| NDRO | Non-destructive readout |
| N-FET | n-type conduction field-effect transistor |
| N-MOS | n-channel metal-oxide silicon |

| | |
|---|---|
| OP-AMP | Operational amplifier |
| OTA | Operational transconductance amplifier |
| PAM | Pulse-amplitude modulation |
| PCCD | Profiled charge-coupled device |
| PCM | Pulse-code modulation |
| PDO | Phosphorous-doped oxide |
| P-FET | p-type conduction field-effect transistor |
| Phot-SCR | Photosensitive silicon controlled rectifier |
| P-JFET | p-type junction-conduction field-effect transistor |
| PLA | Programmable logic array |
| PLL | Phase-locked loop |
| P-MOS | p-type metal-oxide silicon |
| PPM | Pulse position modulation |
| PRF | Pulse repetition frequency |
| PROM | Programmable read-only memory |
| PRR | Pulse repetition rate |
| PSL | Polycrystalline silicon layer |
| PTD | Propagation time delay |
| PTH | Plated-through hole |
| PTM | Pulse time modulation |
| RAM | Random-access memory |
| RAS | Random-access store |
| RCTL | Resistance-coupled transistor logic |
| RF CYCLE | Refresh cycle |
| RLC | Radial lead capacitor |
| ROM | Read-only memory |
| RTL | Resistor–transistor logic |
| SBS | Silicon bilateral switch |
| SCDX | Solid-state control differential transmitter |
| SCR | Silicon-controlled rectifier |
| SDC | Synchro-to-digital converter |
| SEM | Scanning electron microscope |
| Si | Silicon |
| SIC | Silicon integrated circuit |
| Si-gate MOS | Silicon-gate metal-oxide silicon semiconductor |
| SIL | Single-in-line |
| SIP | Single-in-line package |
| SMI | Static memory interface |
| S/N RATIO | Signal-to-noise ratio |
| SOS | Silicon-on-sapphire |
| SPAN | Stored program alphanumeric |
| SR | Status register |
| SRD | Shift register decoder |
| SRT | Shift register transistor |
| SSCT | Solid-state control transformer |
| SSD | Static sensitive device |

| | |
|---|---|
| SSR | Solid-state relay |
| STTL | Schottky transistor–transistor logic |
| TDM | Time-division multiplex |
| TFHC | Thin-film hybrid circuit |
| TFT | Thin-film transistor |
| THB | Through-hole board |
| THP | Through-hole plating |
| TRL | Transistor–resistor logic |
| TTL | Transistor–transistor logic |
| UFET | Unipolar field-effect transistor |
| UJT | Unijunction transistor |
| ULA | Uncommitted logic array |
| VCO | Voltage-controlled oscillator |
| V-FC | Voltage to frequency converter (V/FC) |
| VHSIC | Very-high-speed integrated circuit |
| VMOS | Vertical (structured) metal-oxide semiconductor |
| VSWR | Voltage standing wave ratio |
| VTL | Variable threshold logic |
| XMOS | High-speed metal-oxide semiconductor |

# Appendix 2

## Powers of 2

| $2^n$ | $n$ | $2^{-n}$ |
|---:|:---:|:---|
| 1 | **0** | 1·0 |
| 2 | **1** | 0·5 |
| 4 | **2** | 0·25 |
| 8 | **3** | 0·125 |
| 16 | **4** | 0·062 5 |
| 32 | **5** | 0·031 25 |
| 64 | **6** | 0·015 625 |
| 128 | **7** | 0·007 812 5 |
| 256 | **8** | 0·003 906 25 |
| 512 | **9** | 0·001 953 125 |
| 1 024 | **10** | 0·000 976 562 5 |
| 2 048 | **11** | 0·000 488 281 25 |
| 4 096 | **12** | 0·000 244 140 625 |
| 8 192 | **13** | 0·000 122 070 312 5 |
| 16 384 | **14** | 0·000 061 035 156 25 |
| 32 768 | **15** | 0·000 030 517 578 125 |
| 65 536 | **16** | 0·000 015 258 789 062 5 |
| 131 072 | **17** | 0·000 007 629 394 531 25 |
| 262 144 | **18** | 0·000 003 814 697 265 625 |
| 524 288 | **19** | 0·000 001 907 348 632 812 5 |
| 1 048 576 | **20** | 0·000 000 953 674 316 406 25 |
| 2 097 152 | **21** | 0·000 000 476 837 158 203 125 |
| 4 194 304 | **22** | 0·000 000 238 418 579 101 562 5 |
| 8 388 608 | **23** | 0·000 000 119 209 289 550 781 25 |
| 16 777 216 | **24** | 0·000 000 059 604 644 775 390 625 |
| 33 554 432 | **25** | 0·000 000 029 802 322 387 695 312 5 |
| 67 108 864 | **26** | 0·000 000 014 901 161 193 847 656 25 |
| 134 217 728 | **27** | 0·000 000 007 450 580 596 923 828 125 |
| 268 435 456 | **28** | 0·000 000 003 725 290 298 461 914 062 5 |
| 536 870 912 | **29** | 0·000 000 001 862 645 149 230 957 031 25 |
| 1 073 741 824 | **30** | 0·000 000 000 931 322 574 615 478 515 625 |
| 2 147 483 648 | **31** | 0·000 000 000 465 661 287 307 739 257 812 5 |
| 4 294 967 296 | **32** | 0·000 000 000 232 830 643 653 869 628 906 25 |
| 8 589 934 592 | **33** | 0·000 000 000 116 415 321 826 934 814 453 125 |
| 17 179 869 184 | **34** | 0·000 000 000 058 207 660 913 467 407 226 562 5 |
| 34 359 738 368 | **35** | 0·000 000 000 029 103 830 456 733 703 613 281 25 |
| 68 719 476 736 | **36** | 0·000 000 000 014 551 915 228 366 851 806 640 625 |
| 137 438 953 472 | **37** | 0·000 000 000 007 275 957 614 183 425 903 320 312 5 |
| 274 877 906 944 | **38** | 0·000 000 000 003 637 978 807 091 712 951 660 156 25 |
| 549 755 813 888 | **39** | 0·000 000 000 001 818 989 403 545 856 475 830 078 125 |

# Appendix 3

## Classification of integrated circuits

**Linear**

Amplifiers
  audio
  logarithmic
  operational
  servo
  video
Converters
  analog–digital
  digital–analog
  voltage–frequency
  frequency–voltage
Demodulators
FM detectors
Function generators
Integrators
Level detectors
Mixers
Modulators
Motor speed regulators
Phase-locked loops
Power supply modules
Timers
Voltage comparators
Voltage regulators

**Digital**

Arithmetic logic units
Buffers
Clock generators
Code converters
Comparators
Counters:
  binary
  decade
  storage
Data shifters
Data selectors
Decoders
Demultiplexers
Drivers
  display
  line
Encoders
Flip–flops
Gates
Latches
Line receivers
Memories
Microprocessors
Multiplexers
Multipliers
Multivibrators (monostables)
Parity generators
Pulse synchronizers
Registers
Sense amplifiers
Transceivers
Triggers

# Appendix 4

## Coding of integrated circuit packs

Integrated circuit packs are designed and produced by a variety of specialist manufacturers, and they are available as 'off-the-shelf' units to manufacturers specializing in the design and production of avionic systems. In order to identify units with their particular manufacturer, and also to ensure interchangeability of units, a serial numbering system is adopted. The numbers are prefixed by identifying code letters. Some common 'in-house' codings are as follows.

| | | | |
|---|---|---|---|
| AM | Advanced Micro Devices | PC | General Instrument |
| CA | RCA | RC | Raytheon |
| D | Marconi | RM | Raytheon |
| DM | Signetics | S | Signetics |
| DP | National Semiconductor | SE | Signetics |
| DS | National Semiconductor | SFC | Sescosem |
| H | SGS-Ates | SG | Silicon General |
| HEF | SGS-Ates | SI | Sanken |
| HEF | Mullard | SIG | Signetics |
| ICL | Intersil | SL | Plessey |
| L | SGS-Ates | SN | Texas Instruments |
| LH | National Semiconductor | SP | Signetics |
| LM | Signetics | TAA | SGS-Ates |
| M | Mitsubishi | TRA | SGS-Ates |
| MC | Motorola | TCA | SGS-Ates |
| MFC | Motorola | TDA | SGS-Ates |
| MIC | ITT | TDC | Transitron |
| MM | National Semiconductor | TIL | Texas Instruments |
| MMH | Motorola | TMS | Texas Instruments |
| MP | Micro Power Systems | TOA | Transitron |
| $\mu$A | Fairchild | TVR | Transitron |
| N | Signetics | UHP | Sprague |
| NL | General Instrument | ULN | Sprague |
| NE | Signetics | WC | Westinghouse |
| NH | National Semiconductor | WM | Westinghouse |
| NMC | Newmarket Transistors | ZLA | Ferranti |
| PA | General Electric | | |

All technical and interchangeability data relevant to each coded series of packs is contained in catalogues or data books published by the manufacturers. The packs selected for use in the various circuit modules of avionic systems are identified either by quoting the coding in circuit diagrams, or in the parts lists of systems.

# Appendix 5

## Symbols associated with characteristics of solid-state devices

$BV_{CE}$, $BV_{EC}$, $BV_{CB}$, $BV_R$
Breakdown voltages, collector–emitter, emitter–collector, collector–base, and reverse

$B_W$ Bandwidth

$C_{ISO}$ Isolation capacitance

$C_J$ Junction capacitance

$C_{IN}$ Input capacitance

$C_{I-O}$ Input–output capacitance

CTR Current transfer ratio

$CM$, $CM_H$, $CM_L$, $CM_{RR}$
Common modes, high output, low output and rejection ratio

$G_{BW}$ Gain bandwidth

$I_{CE}$, $I_{CB}$
Leakage currents, collector–emitter and collector–base

$I_F$ Forward current

$I_O$, $I_{OH}$, $I_{OL}$
Output currents, logic high and logic low

$I_{CCH}$, $I_{CCL}$
Supply currents, logic high and logic low

$I_{I-O}$ Input–output current

$I_{EH}$, $I_{EL}$
Enable currents, logic high and logic low

$I_R$ Reverse leakage current

$I_{CC}$ Supply current

$I_{FL}$, $I_{FH}$
Forward currents, low and high

$I_B$ Base photo current

$I_{IL}$, $I_{IH}$
Input currents, low and high

$I_B$, $I_{OS}$
Input currents, bias and offset

$I_{D(off)}$ Drain cut-off current

$I_{D(on)}$ On-state drain current

$I_G$ Gate current

$I_{GF}$, $I_{GR}$
Gate currents, forward and reverse

$I_S$ Source current

$I_n$ Noise current

$N$ Fan out

$PSRR$ Power supply rejection ratio

$R_{I-O}$ Input–output resistance

$R_{ISO}$ Isolation resistance

$SR$ Slew rate

$t_P$ Propagation time

$t_f$ Fall time

$t_r$ Rise time

$T_{PD}$ Propagation delay time

$t_D$ Time delay

$t_{DH}$, $t_{DL}$
Time delay in response to logic high and low inputs

$t_{ON}$ Turn-on time

$t_{OFF}$ Turn-off time

$t_{ELH}$, $t_{EHL}$
Enable propagation delay times, high and low level outputs

$T_A$    Operating temperature

$T_J$    Junction temperature

$V_{ISO}$    Isolation voltage (d.c.)

$V_{CE(sat)}$
Collector–emitter saturation voltage

$V_O, V_{OL}, V_{OH}$
Output voltages, logic low and logic high

$V_{BB}, V_{CC}, V_{EE}$
d.c. supply voltages, base, collector and emitter

$V_{BC}, V_{BE}, V_{CB}, V_{CE}, V_{EB}, V_{EC}$
Voltages, d.c. or average, between base and collector, etc.

$V_R$    Reverse input voltage

$V_{EL}, V_{EH}$
Enable voltages, low and high

$V_{CM}$    Common mode voltage

$V_n$    Noise voltage

$V_{IL}, V_{IH}$
Input voltages, low and high

$V_{GS}$    Gate source voltage

$V_{OS}$    Input offset voltage

# Appendix 6

## Acronyms and abbreviations associated with avionic systems, equipment and controlling signal functions

This Appendix, which is by no means exhaustive, is intended as a guide to the meanings of acronyms and abbreviations found in the documentation dealing with the description, operation, logic signal functions and maintenance of systems and equipment.

| | |
|---|---|
| ACARS | ARINC Communications Addressing and Reporting System |
| ACAS | Airborne Collision and Avoidance System |
| ACCEL | Accelerometer |
| ACQ | Acquire (prefixed by a condition, e.g., ALT ACQ) |
| A/D | Analog to Digital |
| ADC | Air Data Computer |
| ADEU | Automatic Data Entry Unit |
| ADF | Automatic Direction Finder |
| ADI | Attitude Director Indicator |
| AFCS | Automatic Flight Control System |
| AFS | Automatic Flight System |
| AGC | Automatic Gain Control |
| AGS | Automatic Gain Stabilization |
| AHRS | Attitude and Heading Reference System |
| AIDS | Airborne Integrated Data System |
| ALG ARM | Align Arm |
| ALOFT | Airborne Light/Optical Fibre Technology |
| ALPHA | Angle of Attack |
| ALU | Arithmetic Logic Unit |
| ANN | Annunciator |
| AOSS | After Over Station Sensor |
| AP, A/P | Autopilot (suffixed by condition, e.g., ENG, DISC) |
| APFDS | Autopilot and Flight Director System |
| APMS | Automatic Performance and Management System |
| APPR OC | Approach On Course |
| APS | Altitude Preselect |
| APSB | APS Bracket |
| ARINC | Aeronautical Radio InCorporated |
| ARM | Armed (prefixed by condition, e.g., LOC ARM, VOR ARM) |

| | |
|---|---|
| AS, A/S | Airspeed |
| ASA | Autoland Status Annunciator |
| AT | Autothrottle |
| ATC | Air Traffic Control (Altitude reporting) |
| ATE | Automatic Test Equipment |
| ATR | Austin Turnbull Radio (formerly Air Transport Radio) |
| ATS | AutoThrottle System |
| AT/SC | AutoThrottle/Speed Control |
| ATLAS | Abbreviated Test Language for Avionic Systems |
| ATOL | Automatic Test-Oriented Language |
| ATT | Attitude (may be followed by condition, e.g., ATT HOLD) |
| ATT ERR | Attitude Error |
| AUTO APPR | Automatic Approach |
| B/A | Bank Angle |
| BARO | Barometric |
| BB | Bar Bias |
| B/B | Back Beam |
| B/C, BC, B/CRS | Back Course |
| B/D | Bottom of Descent |
| BITE | Built-In Test Equipment |
| BRG | Bearing |
| CAD | Computer-Aided Design |
| CADC | Central Air Data Computer |
| CAP | Capture (prefixed by a condition, e.g., LOC CAP, NAV CAP) |
| CAPS | Collins Adaptive Processor System |
| CAWP | Caution And Warning Panel |
| CBB | Collective Bar Bias |
| CDU | Control and Display Unit |
| CE | Course Error |
| CG | Character Generator |
| CINCH | Compact Inertial Navigation Combining Head-up display |
| CLK | Clock |
| CMD | Command (prefixed by another abbreviation, e.g., FD CMD) |
| COMP | Compensation, Compass, Comparator |
| CONT | Controller |
| CP | Control Panel |
| CPL | Coupled (prefixed by condition, e.g., ROLL, PITCH, APPR) |
| CPU | Central Processor Unit |
| CRS | Course |
| CRT | Cathode Ray Tube |

| | |
|---|---|
| CSEU | Control Systems Electronic Unit |
| CT | Control Transformer |
| CW | Caution and Warning |
| CWS | Control Wheel Steering |
| | |
| D/A | Digital to Analog |
| DAD | Data Acquisition Display |
| DADC | Digital Air Data Computer |
| DAIS | Digital Avionics Information System |
| DDI | Dual Distance Indicator |
| DDS | Digital Display System |
| DEVN | Deviation |
| DES | Desired (suffixed by condition, e.g., DES TRK, DES CRS) |
| DFDAU | Digital Flight Data Acquisition Unit |
| DFDR | Digital Flight Data Recorder |
| DG | Directional Gyroscope |
| DH | Decision Height |
| DI | Digital Interface |
| DIFCS | Digital Integrated Flight Control System |
| DISC | Disconnect |
| DISPL | Displacement |
| DLC | Direct Lift Control |
| DMAP | Digital Modular Avionics Programme |
| DME | Distance Measuring Equipment |
| DMLS | Doppler Microwave Landing System |
| DMM | Digital Multi-Meter |
| DMUX | Demultiplexer |
| DN | Down |
| DP | Differential Protection |
| DRC | Dual Remote Compensator |
| DSB | Double-Sided Board |
| DSR TK | Desired Track |
| DTG | Distance-To-Go |
| DU | Display Unit |
| DVM | Digital Volt-Meter |
| | |
| EADI | Electronic Attitude Direction Indicator |
| ECAM | Electronic Centralized Aircraft Monitor |
| ECS | Environmental Control System |
| EDPS | Electronic Data Processing System |
| EEC | Electronic Engine Control |
| EFCU | Electronic Flight Control Unit |
| EFIS | Electronic Flight Instrument System |
| EGT | Exhaust Gas Temperature |
| EHSI | Electronic Horizontal Situation Indicator |
| EHSV | Electro-Hydraulic Servo Valve |
| EICAS | Engine Indicating and Crew Alerting System |

| | |
|---|---|
| ENG | Engage |
| EO | Easy-On |
| EPR | Engine Pressure Ratio |
| EX LOC | Expanded Localizer |
| EXT | Extend |
| | |
| FAC | Flight Augmentation Computer |
| FADEC | Full Authority Digital Engine Control |
| FAWP | Final Approach Waypoint |
| FCC | Flight Control Computer |
| FCEU | Flight Control Electronic Unit |
| FCU | Flight Control Unit |
| FCES | Flight Control Electronic System |
| FD, F/D | Flight Director |
| FDEP | Flight Data Entry Panel |
| FF | Feeder Fault |
| FGS | Flight Guidance System |
| FIM | Fault Isolation Monitoring |
| FIS | Flight Instrument System |
| FL CH | Flight Level CHange |
| FMA | Flight Mode Annunciator |
| FMC | Flight Management Computer |
| FMCS | Flight Management Computer System |
| FMCU | Flight Management Computer Unit |
| FMS | Flight Management System |
| FODTS | Fiber-Optic Data Transmission System |
| FPC | Fuel Performance Computer |
| FPM | Flap Position Module |
| FS | Fast Slew |
| FSEU | Flap/Slot Electronic Unit |
| FTR | Force Trim Release |
| FVC | Frequency-to-Voltage Converter |
| FWC | Flight Warning Computer |
| | |
| GA, G/A | Go-Around |
| GPWS | Ground Proximity Warning System |
| GS, G/S | Glide Slope |
| | |
| HARS | Heading and Attitude Reference System |
| HDG | Heading (can be suffixed by condition, e.g., HDG HOLD, HDG SELect) |
| HLD | Hold |
| HSI | Horizontal Situation Indicator |
| HUD | Head-Up Display |
| HVPS | High Voltage Power Supply |
| | |
| IAS | Indicated Airspeed |
| IAWP | Initial Approach Waypoint |
| ICU | Instrument Comparator Unit |

| | |
|---|---|
| IDG | Integrated Drive Generator |
| ILS OC | Instrument Landing System On Course |
| IMU | Inertial Measuring Unit |
| INC–DEC | Increase–Decrease |
| INS | Inertial Navigation System |
| INTLK | Interlock |
| INTGL | Integral |
| INWP | Intermediate Waypoint |
| IRMP | Inertial Reference Mode Panel |
| IRS | Inertial Reference System |
| IRU | Inertial Reference Unit |
| ISS | Inertial Sensing System |
| IVS | Instantaneous Vertical Speed |
| IVV | Instantaneous Vertical Velocity |
| LAU | Linear Accelerometer Unit |
| LBS | Lateral Beam Sensor |
| LINS | Laser Inertial Navigation System |
| LNAV | Lateral NAVigation |
| LOC | Localizer |
| LRRA | Low-Range Radar Altimeter |
| LRU | Line Replaceable Unit |
| LSSAS | Longitudinal Static Stability Augmentation System |
| LSU | Logic Switching Unit |
| LVDT | Linear Voltage Differential (also Displacement) Transformer |
| LVPS | Low-Voltage Power Supply |
| MADGE | Microwave Aircraft Digital Guidance Equipment |
| MALU | Mode Annunciation Logic Unit |
| MAN | Manual |
| MAP | Mode Annunciator Panel |
| MAWP | Missed Approach WayPoint |
| MCDP | Maintenance Control Display Panel |
| MCP | Mode Control Panel |
| MCU | Modular Concept Unit |
| MDA | Minimum Descent Altitude |
| MIP | Maintenance Information Printer |
| MM | Middle Marker |
| MPU | Microprocessor Unit |
| MSU | Mode Selector Unit |
| MTP | Maintenance Test Panel |
| MUX | Multiplexer |
| MWS | Master Warning System |
| NAV | Navigation |
| NC | No Connection or Normally Closed |
| NCD | No Computed Data |
| NCU | Navigation Computer Unit |

| | |
|---|---|
| ND | Navigation Display |
| NDB | Non Directional Beacon |
| NOC | NAV On Course |
| | |
| OC, O/C | On Course |
| OD | Out of Detent (may be prefixed, e.g., CWS OD) |
| OF | Over Frequency |
| ONS | Omega Navigation System |
| OM | Outer Marker |
| OSS | Over Station Sensor |
| OV | Over Voltage |
| | |
| PAFAM | Performance And Failure Assessment Monitor |
| PAS | Performance Advisory System |
| P ATT | Pitch Attitude |
| PBB | Pitch Bar Bias |
| PCA | Power Control Actuator |
| PCB | Printed Circuit Board |
| PCPL | Pitch Coupled |
| PCWS | Pitch Control Wheel Steering |
| PDCS | Performance Data Computer System |
| PDU | Pilot's Display Unit |
| PECO | Pitch Erection Cut-Off |
| PFD | Primary Flight Display |
| PIU | Peripheral Interface Unit |
| PMS | Performance Management System |
| PNCS | Performance Navigation Computer System |
| P HOLD | Pitch Hold |
| PRAM | Programmable Analog Module |
| PS | Power Supply |
| PSAS | Pitch Stability Augmentation System |
| PSM | Power Supply Module |
| PSO | Phase Shift Oscillator |
| PSU | Passenger Service Unit also Power Supply Unit |
| P SYNC | Pitch Synchronization |
| PWR | Power |
| | |
| QAR | Quick Access Recorder |
| | |
| RA, R/A | Radio (Radar) Altimeter |
| RALU | Register and Arithmetic Logic Unit |
| RAT | Ram Air Turbine |
| RBA | Radio Bearing Annunciator |
| RBB | Roll Bar Bias |
| RCPL | Roll Coupled |
| RCVR | Receiver |
| RCWS | Roll Control Wheel Steering |
| RDMI | Radio Distance Magnetic Indicator |
| REF | Reference |

| | |
|---|---|
| REV/C | Reverse Course (same as Back Course) |
| RG | Raster Generator |
| R/HOLD | Roll Hold |
| RLS | Remote Light Sensor |
| RMI | Radio Magnetic Indicator |
| RN, RNAV | Area Navigation |
| RN/APPR | Area Navigation Approach |
| RSAS | Roll Stability Augmentation System |
| RSU | Remote Switching Unit |
| R/T | Receiver/Transmitter |
| RTE DATA | Route Data |
| RVDT | Rotary Voltage Differential Transmitter |
| R/W | Read/Write |
| | |
| SAI | Stand-by Attitude Indicator |
| SAM | Stabilizer Aileron Module |
| SAS | Stability Augmentation System |
| SBY | Standby |
| S/C | Step Climb |
| SCAT | Speed Command of Altitude and Thrust |
| SCM | Spoiler/Speedbrake Control Module |
| SEL | Select |
| SELCAL | Selective Calling |
| SFCC | Slat/Flap Control Computer |
| SID | Standard Instrument Departure |
| SG | Symbol Generator (Stroke Generator) |
| SGU | Symbol Generator Unit |
| SPD | Speed (Airspeed or Mach hold) |
| SRP | Selected Reference Point |
| SS | Slow Slew |
| SSEC | Static Source Error Correction |
| STAR | Standard Terminal Arrival Route |
| STS | Status (prefixed by a function, e.g., TRACK STS) |
| STC | Sensitivity Time Control |
| STCM | Stabilizer Trim Control Module |
| | |
| TACAN | Tactical Air Navigation |
| TAS | True Air Speed |
| T/C | Top of Climb |
| TCC | Thrust Control Computer |
| TCS | Touch Control Steering |
| T/D | Top of Descent |
| TET | Turbine Entry Temperature |
| TGT | Turbine Gas Temperature |
| TKE | Track angle Error |
| TK CH | Track Change |
| TMC | Thrust Management Computer |
| TMS | Thrust Management System |

| | |
|---|---|
| TMSP | Thrust Mode Select Panel |
| TRP | Thrust Rating Panel |
| TTG | Time To Go |
| TTL | Tuned To Localizer |
| TURB | Turbulence |
| | |
| UART | Universal Asynchronous Receiver Transmitter |
| UF | Under Frequency |
| UV | Under Voltage |
| | |
| VAR | Variable |
| VBS | Vertical Beam Sensor |
| VDU | Visual Display Unit |
| VGU | Vertical Gyro Unit |
| VLD | Valid (usually suffixing a condition, e.g., VG VLD, FLAG VLD) |
| VNAV | Vertical Navigation |
| VOR | Very-high-frequency Omnidirectional Range |
| VOR APPR | VOR Approach |
| VOR OC | VOR On Course |
| VORTAC | VOR TACtical (Air navigation) |
| VR | Volts Regulated |
| VS | Vertical Speed |
| VSCU | Vertical Signal Conditioner Unit |
| | |
| WO, W/O | Washout |
| WPT | Waypoint |
| WXR | Weather Radar transceiver |
| | |
| XTK DEV | Cross Track Deviation |
| XTR | Transmitter |
| | |
| YD, Y/D | Yaw Damper |
| YDM | Yaw Damper Module |

# Appendix 7

## Circuit diagram symbols

The symbols shown in this Appendix have been selected on the basis of common usage in circuit diagrams and wiring diagrams relevant to aircraft systems and equipment. Some variations may be found when referring to different manufacturer's documentation, but they conform generally to accepted standards.

**Amplifiers**

*General*

*Operational*

**Aural Warning**

**Capacitors**

*General*

*Variable*

Curved element represents the outside electrode in fixed paper and ceramic dielectric capacitors, the negative electrode in electrolytic capacitors, the moving element in variable and the low potential element in feed through capacitors.

**CRT Envelope**

Crystal

**Connectors**

*Separable*

*Test point*

*Plug and receptacle*

*Co-axial (outside conductor carried through)*

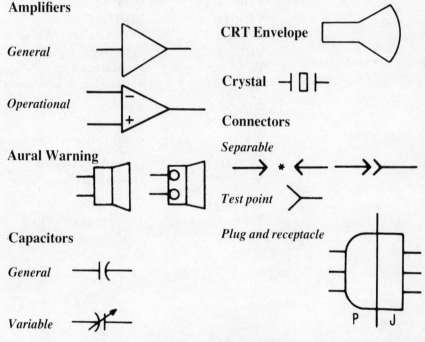

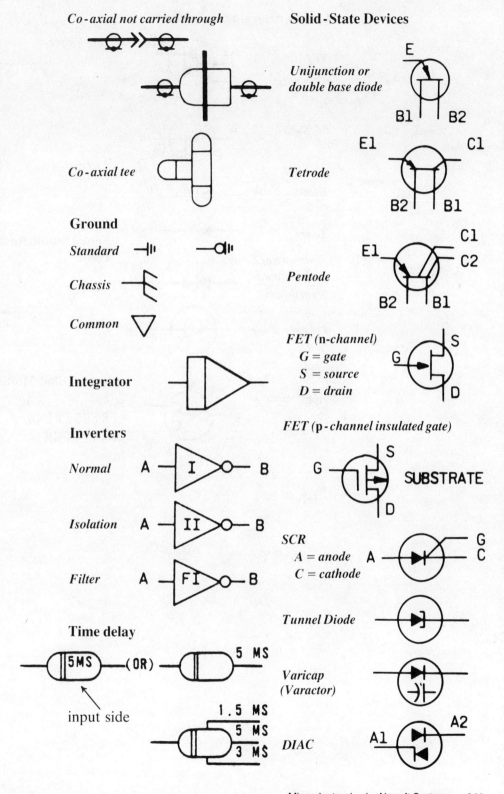

## Circuit diagram symbols–*contd*

*TRIAC*

*SC Switch*

*Diode*

*Zener*

*Symmetrical Zener (bidirectional breakdown)*

*4-Layer diode*

*npn Switch*

*pnp Switch*

*LED*

**Signal Summation Point**

**Single-Shot Multivibrator**

(X) = TRIGGERING WAVEFORM

(Y) = OUTPUT WAVEFORM

# Appendix 8

## Computer languages

**ALGOL** Acronym for ALGOrithmic Language, which is problem-oriented and high-level for mathematical and scientific use. Defines algorithms as a series of statements having a general resemblance to algebraic formulae and English sentences.

**Assembly** An intermediate symbolic language for programming which must go through an assembly for conversion into a machine code.

**BASIC** Acronym for Beginner's All-purpose Symbolic Instruction Code, which is a high-level programming language. A particular feature is that it allows the user to interact with the program while it is being executed. Source language facilities are comparable with FORTRAN.

**COBOL** Acronym for COmmon Business Oriented Language, which is problem-oriented and high-level for general commercial use. The source program is written using statements in English of a standard but readable form.

**FORTRAN** Acronym for FORmula TRANslation language, which is problem-oriented and high-level for scientific, mathematical and engineering use. The source program is written using a combination of algebraic formulae and statements in English of a standard but readable form.

**PASCAL** High-level language used in some Flight Management Systems. Named after the French mathematician and philosopher Blaise Pascal.

# Appendix 9

## Abbreviations and acronyms used in flight management system displays

| | |
|---|---|
| ACT | Active (related to modes) |
| ADC | Air Data Computer |
| ALT | Altitude |
| ALTN | Alternate |
| APPR | Approach |
| APR | Automatic Performance Reserve |
| ARR | Arrive |
| AT | Autothrottle |
| ATIS | Airport Terminal Information Service |
| ATM | Assumed Temperature Method |
| ATO | Actual Time Overhead |
| | |
| BM | Buffet Margin |
| BOD | Bottom Of Descent |
| BRG | Bearing |
| | |
| CG | Centre of Gravity |
| CLB | Climb |
| CMD | Command |
| CON | Continuous (EPR) |
| CRS | Course |
| CRZ | Cruise |
| CSTR | Constraint, e.g., SPD CSTR Speed Constraint |
| CTR | Centred |
| | |
| DA | Drift Angle |
| DECEL | Decelerate |
| DEP | Depart |
| DES | Descent |
| DEST | Destination |
| DEV | Deviation |
| DIR | Direction |
| DIS(T) | Distance |
| DNTK | Downtrack |
| DSPL | Display |
| DSR | Desired |
| DTG | Distance To Go |

| | |
|---|---|
| ECON | Economy |
| E/D | End of Descent |
| EFC | Expect Further Clearance |
| EFOB | Estimated Fuel On Board |
| ELV | Elevation |
| E/OUT | Engine Out |
| EPR | Engine Pressure Ratio |
| ETA | Estimated Time of Arrival |
| ETO | Estimated Time Overhead |
| EXEC | Execute |
| | |
| FF | Fuel Flow |
| FCU | Flight Control Unit |
| FL | Flight Level |
| FLT PLN | Flight Plan |
| FOB | Fuel on Board |
| FOD | Fuel Over Destination |
| FPA | Flight Path Angle |
| FQ | Fuel Quantity |
| | |
| GA | Go Around |
| GS | Ground Speed |
| GW | Gross Weight |
| | |
| IAS | Indicated Air Speed |
| INTC | Intercept |
| IRS | Inertial Reference System |
| ISA | International Standard Atmosphere |
| INIT | Initialization |
| INBD | Inbound |
| INVAL | Invalid |
| | |
| LAT | Lateral |
| LRC | Long Range Cruise |
| LW | Landing Weight |
| | |
| MAG VAR | Magnetic Variation |
| $M_{MO}$ | Maximum operating Mach No. |
| MRC | Mach Referenced Cruise |
| MSA | Minimum Safe Altitude |
| MSG | Message |
| | |
| NM | Nautical Miles |
| $N_1$ | Low pressure compressor speed |
| | |
| OAT | Outside Air Temperature |
| OFST | Offset |
| OPT | Optimum |
| Op Program | Operations Programme |
| | |
| PERF | Performance |

| | |
|---|---|
| PLNG | Planning, e.g., PLNG DES Planning Descent |
| POS | Position |
| PPOS | Present Position |
| PRED | Predicted |
| PROC | Procedure, e.g., PROC TURN Procedure Turn |
| PROF | Profile, e.g., PROF DES Profile Descent |
| PWR | Power |
| | |
| QNH | Altimeter pressure scale setting to read airfield height above sea-level on landing and take-off |
| | |
| RCL | Recall |
| RED | Reduced, e.g., RED TO Reduced take-off thrust |
| RNG | Range |
| RSV | Reserve (fuel) |
| REV | Revision |
| RTE | Route |
| | |
| SAT | Static Air Temperature |
| SEL | Select |
| SID | Standard Instrument Departure |
| SPD | Speed |
| STAR | Standard Terminal Arrival Route |
| STBY | Standby |
| STP | Step, e.g., STEP CLB |
| | |
| TAI | Thermal Anti-Ice |
| TAS | True Air Speed |
| TAT | Total Air Temperature |
| TGT | Target, e.g., TGT ALT Target Altitude |
| TO APR | Take-Off Automatic Performance Reserve |
| TOC (T/C) | Top of Climb |
| TOD (T/D) | Top of Descent |
| TOGW | Take-Off Gross Weight |
| TKE | Track Error |
| TP | Turn Point |
| TRANS | Transition |
| TRK | Track |
| TURB | Turbulence |
| | |
| VAL | Valid |
| VAR | Variation |
| $V_1$ | Critical engine-failure speed |
| $V_2$ | Take-off safety speed |
| VERT | Vertical |
| $V_{MO}$ | Maximum Operating Airspeed |
| $V_{REF}$ | Reference Speed |
| V/S | Vertical Speed |

| | |
|---|---|
| W/V | Wind direction/velocity |
| WPT (WYPT) | Waypoint |
| XTK | Cross Track |
| ZFW | Zero Fuel Weight |

# Appendix 10

## Aircraft utilizing electronic display technology

In addition to the aircraft depicted in the frontispiece and on pages 26 and 46, digital computer and electronic flight instrument display technology are also applied to other types of civil aircraft, of course, and three examples are included in this Appendix. The extent to which flight management data can be presented by means of electronic displays provides for improved efficiency in the operation of the smaller types of transport or business executive aircraft, and the number of applications of these displays to such aircraft rapidly increases. At the present time, the following aircraft in these categories either utilize or have provisions for, the systems stated alongside them:

| | |
|---|---|
| ATR-42 | SFENA IFS 68 (proposed) |
| Beech King Air 200 | Collins 85 EFIS |
| Embraer Brasilia | Collins 85 EFIS (Bendix system optional) |
| Falcon 100 | Collins 85 EFIS (First business executive aircraft to obtain US certification with EFIS) |
| Gulfstream III | Sperry EDZ 800 EFIS |
| Shorts 360 | Collins 85 EFIS |

**Gates Lear Jet 50**—an example of an aircraft utilizing new electronic display technology. By kind permission of CSE Aircraft Services.

**Douglas DC-9 Super-82**—an example of an aircraft utilizing new electronic display technology. By kind permission of Finnair.

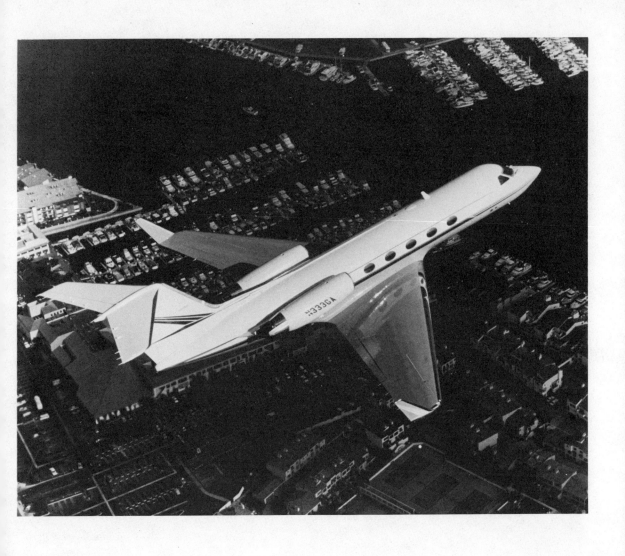

**Gulfstream III**—an example of an aircraft utilizing new electronic display technology. By kind permission of Grumman Aerospace Corporation.

# Bibliography

**Avionics and Equipment**

**Aircraft Electrical Systems,** 2nd edition (Pallett). Pitman.
**Aircraft Instruments,** 2nd edition (Pallett). Pitman.
**Aircraft Radio Systems** (Powell). Pitman.
'An advanced helicopter engine control system.' *RAeS Aeronautical Journal*, November 1982
ARINC Specifications (S) and: Characteristics (C)

| | | |
|---|---|---|
| (S) | 404A | *Air Transport Equipment Cases and Racking* |
| (S) | 408A | *Air Transport Indicator Cases and Mounting* |
| (S) | 410 | *Mark 2 Standard Frequency Selection System* |
| (S) | 542 | *Flight Data Recorders* |
| (S) | 573 | *Digital Flight Data Recorders* |
| (S) | 600 | *Air Transport Avionics Equipment Interfaces* |
| (S) | 616 | *ATLAS Avionics Subset Language* |
| (S) | 701 | *Flight Control Computer* |
| (S) | 702 | *Flight Management Computer System* |
| (S) | 703 | *Thrust Control Computer* |
| (S) | 704 | *Inertial Reference System* |
| (S) | 706 | *Subsonic Air Data Computer* |
| (S) | 707 | *Low Range Radar Altimeter* |
| (S) | 708 | *Airborne Weather Radar* |
| (S) | 709 | *Distance Measuring Equipment* |
| (S) | 717 | *Flight Data Acquisition Systems* |
| (S) | 724 | *Communications Addressing and Recording System* |
| (C) | 725 | *Electronic Flight Instrument Systems (EFIS)* |
| (S) | 726 | *Flight Warning Computer System* |
| (S) | 727 | *Microwave Landing System* |

**Automatic Flight Control,** 2nd edition (Pallett). Granada Publishing.
**Aviation Electronics,** 4th edition (Bose). Prentice Hall International.
'Avionics–the third dimension' (Hearne). *RAeS 'Aerospace'*, March 1981.
**Electricity and Electronics for Aerospace Vehicles,** 2nd edition (McKinley and Brent). McGraw-Hill.
'Electronics, aeronautics and space' (Puckett). *RAeS 'Aerospace'*, February 1981.
ICAO Annex 10–Aeronautical Telecommunications
'Integrated aircraft avionics and powerplant control and management systems.' *RAeS Aeronautical Journal*, December 1982.

**Manual of Avionics** (Kendal). Granada Publishing. (Deals principally with air traffic control aspects.)

**Computers**

'Computers on the airliner flight deck' (Hirst). *Flight International*, 31 March 1979.
'Automation; servant or master?' (Williams and Dugan). *Flight International*, 25 April 1981.
**Electronic Computers Made Simple** (Jacobowitz). W. H. Allen.
**Electronic Computers** (Hollingdale and Tootill). Penguin Books.
**Introduction to Digital Computer Technology** (Nashelsky). John Wiley.
**Penguin Dictionary of Computers** (Chandor). Penguin Books.

**Digital Techniques, Data Transfer and Logic**

(see also Symbols and Diagrams)

ARINC
    (S) 419   *Digital Data System.*
    (S) 429   *Digital Information Transfer System (DITS).*
    (S) 453   *Very High Speed Data Bus.*
**Beginners Guide to Digital Techniques** (Rubasoe). Butterworth.
**Beginners Guide to Digital Electronics** (Sinclair). Butterworth.
**Digital Systems—Principles and Applications** (Tocci). Prentice Hall International.
**Digital Techniques, Books 1 and 2** (Heathkit Learning Publications). Heath Company, Michigan, USA.
**Digital Techniques and Systems** (Green). Pitman.
**Electronic Logic Circuits** (Gibson). Edward Arnold.
**Understanding Digital Electronics.** Texas Instruments Ltd, Bedford, UK.

**Displays**

'Airborne electronic displays' (Hunt). *Proceedings of the Institution of Electrical Engineers*, 1981, Vol. 128, Part A, No. 4
**Alphanumeric Displays, Devices, Drive Circuits and Applications** (Weston and Bittleston). Granada Publishing.
ARINC (S)589: *CRT Displays*
ARINC (S)601: *Control/Display Interface*
'Evolution of the airborne display' (Stonehouse). *Electronic Engineering*, mid-October, 1980.
'Investigation into the optimum use of advanced displays in future transport aircraft. (Hillman & Wilson). *RAeS Aeronautical Journal*, September 1976.
**Optoelectronics—Theory and Practice.** Texas Instruments Ltd.
**Understanding Optronics** (Master). Texas Instruments Ltd.

**Electronics**

ARINC (S)409A: *Selection and Application of Semiconductor Devices.*
**Beginner's Guide to Transistors** (Reddihough). Butterworth.

**Electronics,** II (Green). Pitman.
**Electronic Testing and Fault Diagnosis** (Loveday). Pitman.
**Electronic Fundamentals and Applications** (Ryder). Pitman.
**Elements of Electronics,** Book 3, **Semiconductor Technology** (Wilson). Bernard Babani (Publishing).
**Handbook for Electronic Engineering Technicians** (Kaufman and Seidman). McGraw-Hill.
**Operational Amplifiers, their Principles and Applications** (Dance). Butterworth.
**Operational Amplifiers (Application Manual B169)** (R. Mann). Texas Instruments Ltd.
**The New Penguin Dictionary of Electronics** (Young). Penguin Books.
**Understanding Solid-state Electronics.** Texas Instruments Ltd.
**Applications of Field-effect Transistors (TP1762).** Mullard Ltd.

## Equipment Handling

BS 5783: *Code of Practice for the Handling of Electrostatic Sensitive Devices.* British Standards Institution.
DEF/STAN 59/98: *Handling Procedures for Static Sensitive Devices.*
IPC-T-50: *Terms and Definitions* (*Interconnecting and Packaging*). Institute for Interconnecting and Packaging Electronic Circuits, USA.

## Glossaries and Terminology

BS 3527: *Glossary of Terms used in Data Processing.* British Standards Institution.
MIL-STD-429: *Printed Wiring and Printed Circuit Terms and Definitions.*

## Integrated Circuits

**Beginners Guide to Integrated Circuits** (Sinclair). Butterworth.
**MOS Integrated Circuits and their Applications.** Mullard Ltd.
**Questions and Answers on Integrated Circuits** (Hibbard). Butterworth.
**TTL: from the beginning Application Report B124** (D. A. Bonham). Texas Instruments Ltd.

## Microprocessors

**Basic Principles and Practice of Microprocessors** (Heffer, King and Keith). Edward Arnold.
**Elements of Electronics,** Book 4, **Microprocessing Systems and Circuits** (Wilson). Bernard Babani (Publishing).
**Fundamentals of Microprocessor Design.** Texas Instruments Ltd.
**Introducing Microprocessors** (Sinclair). Keith Dickson Publishing.
**Introduction to Microprocessors** (Aspinall and Douglas). Pitman.
**Microprocessor Fundamentals** (Halsall and Lister). Pitman.
**Microprocessors, a Short Introduction** (Morgan). UK Department of Industry.
**Microprocessors Theory and Applications** (Strutmatter and Fiore). Reston.
**The Penguin Dictionary of Microprocessors** (Chandor). Penguin Books.

**Understanding Microprocessors.** Texas Instruments Ltd.
**An Introduction to Microprocessors (Application Report B190).** Texas Instruments Ltd.
**Microprocessor Serial Code Generation (Application Report B182)** (C. Gase). Texas Instruments Ltd.

**Printed Circuits**

(see also Glossaries & Terminology)
*The Design and Drafting of Printed Circuits* (Darryl Lindsey). Bishop Graphics Inc.

**Symbols and Diagrams**

American National Standards Institute:
    ANSI Y14–15–1970: *Electrical and Electronic Diagrams*
    ANSI Y32–2–1970: *Graphical Symbols for Electrical and Electronic Diagrams*
    ANSI Y32–14–1973: *Graphical Symbols for Logic Diagrams*
    ASA–Y32–14: *American Standard Graphical Symbols for Logic Diagrams*
British Standards Institute
    BS 3363: *Letter Symbols for Semiconductor Devices*
    BS 3939: *Graphical Symbols for Electrical and Electronic Diagrams*
Institute of Electrical & Electronic Engineers (USA)
    IEEE 315–1971: *Graphical Symbols for Electrical and Electronic Diagrams*
    IEEE Std. 91–1973: *Graphic Symbols for Logic Diagrams*

# Exercises

**Chapter 1**

1. What are the three basic configurations of printed circuit boards?
2. For what purpose is a plated-through hole required?
3. What processes are involved in the printing of circuits?
4. Explain the differences between thin and thick films.
5. Why is a monolithic integrated circuit so called?
6. What is meant by the term substrate in connection with integrated circuits?
7. How is packing density normally defined?
8. How is it ensured that the elements of the various active components of a monolithic integrated circuit are accurately diffused within the substrate?
9. Name the types of integrated circuit pack in current use.
10. How does the fabrication process of a hybrid integrated circuit differ from that adopted for a monolithic?

**Chapter 2**

1. Determine the decimal values of the following numbers:
   (a) $10110_{(2)}$,
   (b) $110111_{(2)}$.
2. Convert the following numbers into binary:
   (a) $35_{(10)}$,
   (b) $127_{(10)}$.
3. Convert $142_{(8)}$ into decimal.
4. $139_{(10)}$ is equal to . . . $_{(8)}$?
5. What is the largest decimal number that can be represented with 8 bits?
6. Convert the following numbers into hexadecimal:
   (a) $28_{(10)}$,
   (b) $192_{(10)}$,
   (c) $249_{(10)}$.
7. Determine the decimal numbers of the following:
   (a) $2F_{(16)}$,
   (b) $F5_{(16)}$,
   (c) $A6_{(16)}$.
8. Convert the following into hexadecimal:
   (a) $1011101010010010_{(2)}$,
   (b) $1111000100001001_{(2)}$.

9   Determine the binary numbers of the following:
    (a) $A26F_{(16)}$,
    (b) $2C0_{(16)}$.
10  What is the bit position weighting of a BCD number?
11  From the bit position weighting referred to in Question 10, write the following numbers in BCD code, and show how the numbers obtained compare with pure binary coded numbers:
    (a) $141_{(10)}$,
    (b) $353_{(10)}$,
    (c) $2179_{(10)}$.
12  Write the following in XS3 code:
    (a) $25_{(10)}$,
    (b) $629_{(10)}$,
    (c) $3271_{(10)}$.
13  Convert the following numbers in the indicated code forms into decimal:
    (a) 5-4-2-1 code 1011 1100 0100,
    (b) 7-4-2-1 code 1010 0001 0110.
14  What is the particular property of the 2-out of-5 code?
15  Express $285_{(10)}$ in the following codes:
    (a) 2-out of-5,
    (b) 51111.
16  For what purpose is the Gray code used, and how does it differ from pure binary?
17  For what purpose is the ASCII code used? State the format of a code word.
18  What is the ASCII code for the letter G?
19  Determine whether or not there is an error in the following parity-coded BCD words:

| Word | Parity bit | Parity type |
|---|---|---|
| (a) 1001 | 0 | Odd |
| (b) 1000 | 0 | Odd |
| (c) 0001 | 0 | Even |
| (d) 0110 | 1 | Odd |

20  The following voltage levels appear on six parallel data lines designated A to F:

    A: +5 V;  B: +5 V;  C: 0 V;  D: +5 V;  E: 0 V;  F: +5 V.

    Using positive logic and assuming bit A is the LSB, what is the decimal number equivalent?
21  Add the binary equivalents of $23_{(10)}$ and $13_{(10)}$.
22  Subtract 111010 from 1011111 and state the decimal equivalent.
23  What is meant by the 'ones complement' and the 'twos complement' of a binary number?
24  By means of the ones complement method, subtract 01011 from 11101.

25 Multiply the binary equivalent of $27_{(10)}$ by that of $21_{(10)}$.
26 Divide 11011 by 01001.

# Chapter 3

1 Name the type of gate represented by the symbol shown in Fig. E.1, and list the digital logic notation corresponding to the voltage levels at inputs A and B.

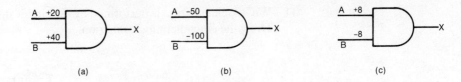

**Fig E.1**

    (a)            (b)            (c)

2 To which logic gate does the following truth table correspond?

| A | B | X |
|---|---|---|
| 1 | 1 | 1 |
| 1 | 0 | 1 |
| 0 | 1 | 1 |
| 0 | 0 | 0 |

3 Identify the logic gate shown in Fig. E.2 and compile its corresponding truth table.

**Fig E.2**

4 What is the significance of a line drawn over a letter or signal function in a logic expression?
5 In compiling a truth table for a 5-input logic gate, how many columns and rows are required to contain the digital logic notation?
6 Compile the truth table for a 4-input OR gate.
7 Name the functions performed by the three basic logic gates.
8 In what way do the symbols signifying logical operation in a Boolean equation differ from those used in a conventional algebraic equation?
9 Figure E.3 represents a very simple fuel pump system. Show how this can also be illustrated by an equivalent logic symbol.

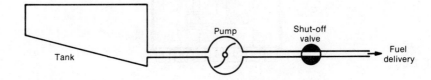

**Fig E.3**

10   In logic terms, what is the function of the air supply system shown in Fig. E.4? Draw the logic equivalent of the system

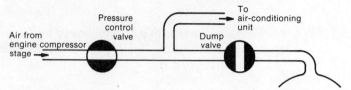

**Fig E.4**

11   What logic function is performed by the circuit shown in Fig. E.5? Draw the equivalent logic diagram.

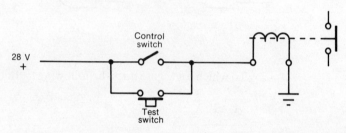

**Fig E.5**

12   When the voltage that represents a logic 1 is less than the voltage that represents a logic 0, the logic is said to be:
(a) inhibited,
(b) positive,
(c) negative.

13   When a 'state indicator' symbol is drawn at the input side of an inverter, it means that for the input signal to be an active one it:
(a) must only be high,
(b) can be either high or low,
(c) must only be low.

14   The equation $X = \overline{AB} + AB$ relates to:
(a) a NAND gate,
(b) an exclusive NOR gate,
(c) an exclusive OR gate.

15   What is the significance of a Karnaugh map?

16   How are the functions of AND and OR logic gates changed in respect of the logic levels assigned to them?

17   Explain how NAND and NOR gates can be made to perform AND and OR logic functions.

18   What logic circuit arrangement is contained within the DIL IC pack shown in Fig. E.6?

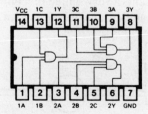

**Fig E.6**

260   Exercises

**Chapter 4**

1. What is the basic function of a flip–flop?
2. What do you understand by the term complement output and how is it indicated on the symbol for a flip–flop?
3. For what purpose are clocked flip–flops used?
4. Describe the operation of a master–slave flip–flop arrangement.
5. The notation —o▷ on a flip–flop logic symbol indicates a:
   (a) 'clear' input,
   (b) 'clocked' input,
   (c) 'toggled' input.
6. What are the functions of 'preset' and 'clear' inputs to some types of flip–flop?
7. What functions can be performed by a shift register?
8. Assuming that the binary word 0011 is stored in a shift register, show by means of a block diagram how the word 1101 generated as a serial input would be shifted in.
9. Explain the difference between a static type and a dynamic type of MOS register.
10. For clocking purposes, rectangular output pulses of specific frequency are generated by:
    (a) a monostable multivibrator,
    (b) an astable multivibrator,
    (c) a frequency regulator unit.
11. Explain the purpose of multiplexing, and state some applications of a multiplexer.
12. What are the basic operating modes of a multiplexer?
13. Why is an exclusive OR gate used in a binary adder circuit?
14. How are the carry functions of binary adders accomplished?
15. In which applications are linear operating range op-amps used?
16. Under what conditions would the polarity of an op-amp output signal be opposite to that of its input signal?
17. How is the feedback component of an op-amp connected to produce negative feedback?
18. An op-amp in which the output voltage is fed back directly to an input performs the function of:
    (a) non-inverting,
    (b) integrating,
    (c) voltage-following.
19. With the aid of a diagram, explain the operation of a summing amplifier.
20. What stable states of an op-amp bistable multivibrator are equivalent to those in a logic gate flip–flop, and how are they changed?
21. What do you understand by the terms 'offset' and 'slewing rate'?

**Chapter 5**

1. What is the difference between an active display and a passive display?
2. By means of a diagram show the construction of a 7-segment LED display.

3  What segment configurations are used for the display of alphanumeric data?
4  How does a dot-matrix LED display differ from a segmented type?
5  How are segments formed in an LCD?
6  Is an LCD of the active or passive type?
7  Explain how a dynamic-scattering type of LCD produces (a) a transmissive readout, and (b) a reflective readout.
8  What differences occur between the appearance of digits and background of dynamic-scattering and field-effect type LCDs when used in the transmissive and reflective readout configurations?
9  How is binary coded decimal information decoded to provide a segmented display?
10  With the aid of Fig. 5.16, determine the decimal number that would be displayed for the encoded binary number 1000100000.
11  In a binary counter using J–K flip–flops, the counter state will change when the toggle input changes from:
   (a) 1 to 0,
   (b) 0 to 1,
   (c) both (a) and (b).
12  A 4-bit binary counter contains the number 0100. After nine input pulses have been applied, the new counter state will be:
   (a) 0010,
   (b) 1011,
   (c) 1101.
13  A 4-bit binary counter contains the number 1011. After eight input pulses have been applied, the new counter state will be:
   (a) 0101,
   (b) 0011,
   (c) 1111.
14  How does a binary down-counter differ from an up-counter?
15  A 4-bit binary down-counter is in the 0110 state. What would its new output state be after the application of 14 input pulses?
16  In terms of output and input connections, how does a binary down-counter differ from an up-counter?
17  How can up- and down-counting capabilities be combined within a single counter?
18  What circuit changes are necessary for a counter to count in the BCD sequence?
19  After a BCD counter has been reset, the number of applied input pulses for a 0111 state would be:
   (a) 12,
   (b) 3,
   (c) 7.
20  What is the purpose of connecting BCD counters in cascade?
21  In a cascaded counter arrangement, which counter represents the MSD?
22  Which of the four outputs of BCD counters in cascade is the LSB?

23  What decimal number would be counted when the binary outputs from three counters in cascade sequence 1, 2, 3 are respectively 0100 1010 0110?

24  Explain how a counter is arranged in order to achieve synchronized flip–flop operation.

25  If the input frequency to a 12-bit counter is 200 kHz, the output frequency will be:
(a) 12·5 kHz,
(b) 0·0488 kHz,
(c) 0·0030 kHz.

## Chapter 6

1  What are the principal operating elements of a cathode ray tube?
2  Describe how the electron beam of a cathode ray tube is made to trace out a display on the screen.
3  How is the electron beam focused?
4  What is the difference between rho–theta and raster scanning?
5  What scanning method is adopted in weather radar indicators which can also display alphanumeric data?
6  How many electron guns are provided in a colour cathode ray tube to produce all the colours required?
7  Explain how the electron beam from a blue gun produces that colour on the tube screen.
8  How is colour mixing obtained?
9  What digital circuit elements are used in a weather radar indicator to produce the output words corresponding to the colours to be displayed?
10  To which level of return echoes and prevailing weather conditions do the binary data 0 1 and 1 1 correspond?
11  The colour corresponding to the binary data level 1 0 is:
(a) red,
(b) green,
(c) yellow.
12  Briefly explain how alphanumeric data are displayed.
13  Define the acronyms EFIS, EICAS and ECAM.
14  What are the principal units that comprise an EFIS installation?
15  How many card modules are provided in each display unit? State the function of each.
16  For what displays is raster scanning used in both display units of EFIS?
17  What is the purpose of a remote light sensor?
18  Which system inputs are supplied to the left symbol generator unit?
19  Explain how radio altitude is displayed on the ADI, and the changes which take place in the display below 1000 feet.
20  Any deviations beyond normal ILS glide slope and/or localizer limits are indicated by the respective deviation pointers on the ADI changing colour from:
(a) green to flashing red,
(b) white to flashing amber,
(c) white to flashing red.

21  What information is displayed on the HSI in either the MAP or PLAN Modes?
22  What information is provided by the curved track vector and the range to altitude arc displayed on the HSI?
23  In which modes is the expanded compass rose displayed?
24  What are the principal units that comprise an EICAS installation?
25  What information is displayed on the upper and lower display units?
26  How are values of the parameters $N_1$ and $EGT$ which exceed normal, displayed and recorded?
27  For what purpose are the ENG and STATUS switches on the EICAS control panel used?
28  Under what conditions is the compacted mode of display used?
29  What is the purpose of the standby engine indicators, and to which parameters do they relate?
30  What is meant by 'auto event' and 'manual event', and how is associated data called up for display?
31  In respect of display format, how does the ECAM system differ from EICAS?
32  Which of the display modes are automatically selected?
33  What displays are presented in the manual mode and how are they selected?
34  Which mode is used in normal operation of the ECAM system?
35  In connection with the modes referred to in Q.34, what displays are presented?
36  In relation to warning displays, what is the significance of a system title which is underlined, and one which is boxed?
37  What information is provided by a STATUS message display?
38  From which units of the ECAM system can self-testing be carried out?
39  How are failures of an FWC and a symbol generator annunciated?
40  What checks are performed during a manual self-test routine?

## Chapter 7

1  Identify the matrix diagram in Fig. E.7, and determine the combined equation for the output F.

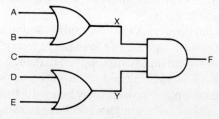

Fig E.7

2  Figure E.8 illustrates a logic gate combination forming part of a pitch-axis logic gate circuit of an automatic flight control system computer. From the input signal data given, determine the combined equation for the output to the pitch axis control section.

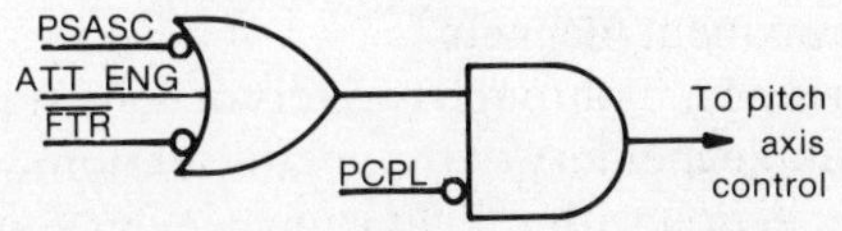

**Fig E.8**

[Diagram: PSASC, ATT ENG, FTR inputs to NOR gate; PCPL to NAND gate; output "To pitch axis control"]

3  With reference to Fig. 7.6, define the signal function abbreviations AOSS, $\overline{\text{NOC}}$, and NOCEO.

**Chapter 8**

1   What do you understand by the analog concept of measurement and computation?
2   What are the elements that constitute the hardware and software of a digital computer?
3   Name the principal sections of a CPU and explain their functions.
4   Briefly explain the functions of the busses comprising a computer highway.
5   Which of the busses referred to in Q.4 are bidirectional?
6   Explain the difference between interrupt and DMA functions.
7   A hybrid computer is one which:
    (a) contains mainly hybrid ICs,
    (b) processes data in both analog and digital forms,
    (c) handles only restricted type programs.
8   What is the name given to the digital code through which a computer carries out instructions?
9   An assembly language program is one by which instructions:
    (a) involve a compiler,
    (b) are given in a mnemonic code,
    (c) are directly in machine code.
10  In the ARINC 429 format of data transfer, how is data identified according to function?
11  What is the reason for assigning two bits of a coded output to identification of source and destination?
12  How is binary data encoded by a data bus?
13  What is meant by the access time of a memory, and in what unit is it measured?
14  A 16K memory has a bit storage capacity of:
    (a) 16 000,
    (b) 32 000,
    (c) 16 384.
15  Explain the difference between a dynamic and a static RAM.
16  Describe the matrix arrangement of a memory.
17  Briefly describe how a stored word in a ROM is read out.
18  What techniques are adopted to permit changes to be made to the program stored in a ROM?
19  What materials are used for the thin film and substrate of a bubble memory?
20  The bubbles are generated in a bubble memory by:
    (a) passing current pulses through aluminium conductor loops,

(b) permanent magnets,
(c) passing a.c. through two sets of coils at right angles to each other.
21 Is a bubble memory of the volatile or non-volatile type?
22 In connection with a flight management system, what is the difference between advisory and combined functions?
23 What essential data inputs are required for operating a flight management system in a full three-dimensional capacity?
24 What are the principal components of a PDC system?
25 What do the three symbols displayed in the control and display unit of a PDC system signify?
26 State the purpose of the command 'bugs' of Mach/airspeed and EPR indicators, and also the methods of coupling them to the computer of a PDC system.
27 How are the various pages of a set called up for display?
28 Briefly explain how computer-requested data is entered into the display.
29 How are the data base records of a flight management computer system initially programmed and transferred to the computer memory?
30 Describe the configuration of an FMCS installation.
31 What is the function of the EXEC key on the display unit, and at what stage does the bar illuminate?
32 What is the purpose of the line keys on either side of a display unit?
33 For what purpose is the bottom line of the display unit used?
34 Explain how data may be changed by over-writing new values.
35 What is the sequence of constructing a flight plan?
36 At which stages of a flight are the CLB and CRZ annunciators illuminated?

## Chapter 9

1 Which operating modes of an ATC SSR system are used for identification and altitude reporting?
2 How does an airborne transponder 'recognize' the mode in which it is being interrogated?
3 Assuming that for identification purposes the code 4510 had been selected, which reply pulse would make up the pulse train?
4 Name the binary code used in an encoding altimeter, and briefly describe how reply pulses are transmitted.
5 What parameters are mandatory for crash investigation purposes?
6 How are the encoded signals appropriate to a parameter 'written' on the recording medium of an electromagnetic type of flight data recorder?
7 Briefly explain the operation of a VOR system.

# Solutions to Exercises

**Chapter 2**

1. (a) $10110_{(2)} = 22_{(10)}$,
   (b) $110111_{(2)} = 55_{(10)}$.

2. (a) $35_{(10)} = 100011_{(2)}$,
   (b) $127_{(10)} = 1111111_{(2)}$.

3. $142_{(8)} = (1 \times 8^2) + (4 \times 8^1) + (2 \times 8^0)$
   $= 64 + 32 + 2$
   $= 98_{(10)}$.

4. $139_{(10)} = 213_{(8)}$.

5. $2^8 - 1 = 256 - 1 = 255_{(10)}$.

6. (a) $28_{(10)} = 1C_{(16)}$,
   (b) $192_{(10)} = C0_{(16)}$,
   (c) $249_{(10)} = F9_{(16)}$.

7. (a) $47_{(10)}$,
   (b) $245_{(10)}$,
   (c) $166_{(10)}$.

8. (a) $BA92_{(16)}$,
   (b) $F109_{(16)}$.

9. (a) 1010 0010 0110 1111,
   (b) 0010 1100 1001.

10. The bit position weighting of a BCD number is $2^3\ 2^2\ 2^1\ 2^0$ or 8-4-2-1.

11. (a) $141_{(10)} = 0001\ 0100\ 0001$ in BCD and $10001101_{(2)}$,
    (b) $353_{(10)} = 0011\ 0101\ 0011$ in BCD and $101100001_{(2)}$,
    (c) $2179_{(10)} = 0010\ 0001\ 0111\ 1001$ in BCD and $100010000011_{(2)}$.

12. (a) $25_{(10)} = 0101\ 1000$,
    (b) $629_{(10)} = 1001\ 0101\ 1100$,
    (c) $3271_{(10)} = 0110\ 0101\ 1010\ 0100$.

13  (a) 894,
    (b) 916.

14  The 2-out-of-5 code has the particular property that there are only two bit 1s in each code group.

15  $285_{(10)}$ = (a) 00110 10100 01100,
              (b) 00011 11110 10000.

16  The Gray code is used with optical or mechanical shaft position encoders, and differs from the pure binary in that only one bit changes between any two successive words.

18  The code for the letter G is 1000111.

19  (a) and (c) are incorrect; (b) and (d) are correct.

20  FEDCBA = 101011 = $43_{(10)}$.

21  The binary equivalent of $23_{(10)}$ is 10111 and of $13_{(10)}$ 01101; therefore:

$$\begin{array}{r} 10111 \\ +01101 \\ \hline 100100 \end{array} = 36_{(10)}.$$

22  $\begin{array}{r} 1011111 = 95 \\ -111010 = 58 \\ \hline 100101 = 37_{(10)}. \end{array}$

24  $\begin{array}{r} 11101 = 29 \\ -01011 = 11 \end{array} = \begin{array}{r} 11101 \\ +10100 \\ \hline 110001 \end{array}$ (ones complement)

The positive sign digit (1) is shifted round to the right-hand side of the number, and added to the digit already there. The difference is therefore 10010 = $18_{(10)}$.

25  The binary equivalents of $27_{(10)}$ and $21_{(10)}$ are respectively 11011 and 10101; therefore;

$$\begin{array}{r} 11011 \\ \times 10101 \\ \hline 11011 \\ 00000\phantom{0} \\ 11011\phantom{00} \\ 00000\phantom{000} \\ 11011\phantom{0000} \\ \hline 1000110111 \end{array} = 567_{(10)}.$$

```
        26         11
   01001)11011
              1001
              ————
              1001
              1001
              ————
              0000
```

# Chapter 3

1. The symbol represents an AND gate, and the corresponding logic notations are as follows:

   (a) A logic 0,
       B logic 1;
   (b) A logic 1,
       B logic 0;
   (c) A logic 1,
       B logic 0.

2. An OR gate.

3. The logic gate shown is an inhibited or negated OR gate and its truth table is constructed as follows:

   Since there are two inputs there will be three columns including the output. From the expression $2^n$ there will be four rows. The columns are identified as $2^0$, $2^1$ starting at the input B column, and by raising 2 to its appropriate power, the alternate placing of the 0s and 1s in the four rows will be the same as for the basic OR gate. Input B is, however, negated so that the outputs from three of the combinations will be opposite to those of the basic OR gate. The table is therefore:

   | $2^1$ | $2^0$ | |
   |---|---|---|
   | A | $\bar{B}$ | X |
   | 0 | 0 | 1 |
   | 0 | 1 | 0 |
   | 1 | 0 | 1 |
   | 1 | 1 | 1 |

5. For a 5-input logic gate the digital logic notation would be contained within six columns (including the output) and $2^5$ or 32 rows.

6. Since there are four inputs there will be four columns plus one for the output. From the expression $2^n$ there will be $2^4$ or 16 rows. From the right the columns are identified as $2^0$, $2^1$, $2^2$, and $2^3$, and by raising 2 to its appropriate power, the alternate placing of the 0s and 1s in the four columns, and therefore the sixteen combinations, will be as shown in the table.

| $2^3$ | $2^2$ | $2^1$ | $2^0$ | |
|---|---|---|---|---|
| A | B | C | D | X |
| 0 | 0 | 0 | 0 | 0 |
| 0 | 0 | 0 | 1 | 1 |
| 0 | 0 | 1 | 0 | 1 |
| 0 | 0 | 1 | 1 | 1 |
| 0 | 1 | 0 | 0 | 1 |
| 0 | 1 | 0 | 1 | 1 |
| 0 | 1 | 1 | 0 | 1 |
| 0 | 1 | 1 | 1 | 1 |
| 1 | 0 | 0 | 0 | 1 |
| 1 | 0 | 0 | 1 | 1 |
| 1 | 0 | 1 | 0 | 1 |
| 1 | 0 | 1 | 1 | 1 |
| 1 | 1 | 0 | 0 | 1 |
| 1 | 1 | 0 | 1 | 1 |
| 1 | 1 | 1 | 0 | 1 |
| 1 | 1 | 1 | 1 | 1 |

9   Since the system requires a tank to house the fuel, a pump and a valve, and because all three must be in an operative state to obtain fuel delivery, the system performs the function of a 3-input AND gate. Its equivalent logic symbol would therefore be as shown in Fig. S.1.

**Fig S.1**

10  The air supply is required for the primary purpose of pressurizing the cabin of an aircraft, and to ensure this the pressure control valve must be open and the dump valve must remain closed. For pressurized conditions, therefore, the dump valve performs a NOT function in logic terms. Since it must also be possible to 'dump' the air, the system overall is equivalent to an inhibited or negated AND gate, as shown in Fig. S.2.

**Fig S.2**

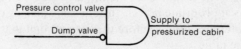

11  The relay may be energized by closing either the control switch or the test switch. The circuit therefore performs an OR function, as shown in Fig. S.3.

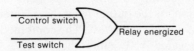

Fig S.3

12  (c).
13  (c).
14  (b).
18  The diagram shows a triple 3-input AND circuit arrangement.

**Chapter 4**

5   (b).
10  (b).
18  (c).

**Chapter 5**

10  The encoded number 1000100000 would be converted by the nine NAND gates to 111011111 and then transmitted to the four NAND gates, whose output in BCD would be 0100. After decoding by the BCD-7-segment decoder, the segments b, c, f and g would be illuminated, and so the decimal number 4 would be displayed.
11  (a).
12  (c) Since 0100 is the binary equivalent of $4_{(10)}$, then after nine input pulses, the total applied will be 13, which has the binary equivalent of 1101.
13  (b) 1011 is the binary equivalent of $11_{(10)}$ and is already in the counter. Since the maximum count capability of a 4-bit counter is 15, it will in this case reach maximum count after the fourth input pulse is applied. The fifth input pulse will recycle the counter to 0000. Therefore on the eighth pulse the new counter state will be 0011, which is the binary equivalent of $3_{(10)}$.
15  The existing 0110 state is equivalent to $6_{(10)}$. Thus, after the first six input pulses the counter would be decremented to 0000. On the seventh input pulse the counter recycles to 1111 ($15_{(10)}$). After the next seven pulses the counter would therefore be decremented to the new counter state 1000 ($8_{(10)}$).
19  (c).
23  Binary 0100 is $2_{(10)}$, binary 1010 is $5_{(10)}$ and binary 0110 is $6_{(10)}$. Since the output from counter 1 is the LSD, and the output from counter 3 the MSD, then the number counted is $652_{(10)}$.
25  (b).

**Chapter 6**

11  (c).
20  (b).

**Chapter 7**

1. The diagram represents a complex AND matrix. The output from OR gate 1 is A + B = X, and the output from OR gate 2 is D + E = Y. Therefore, the output from the AND gate is represented by the combined equation $(A + B) \cdot C \cdot (D + E) = F$.
2. The equation for the output from the negated OR gate is: $\overline{ATT\ ENG + FTR + \overline{PSASC}}$. The combined equation for the output from the negated AND gate is therefore, $(\overline{ATT\ ENG + FTR + \overline{PSASC}}) \cdot (\overline{PCPL})$.

**Chapter 8**

7. (b).
9. (b).
14. (c).
20. (a).

# Index

Access time, 156
Active display, 81
Additive process, 8
Address, 151
Address bus, 151
Addressable locations, 157
ADI display, 121
AFCS annunciator system, 142
ALGOL, 243
All-purpose computer, 152
Alphanumeric
  address generator, 116
  code, 38
  data, 113
  display, 81, 86, 115
Altitude reporting system, 189
Analog computer, 149
Analog-to-digital converter, 153, 163, 167, 207
AND gate, 49
AND matrix, 137
ARINC 429 format, 155
Arithmetic logic unit, 150
ASCII code, 39
Assembler program, 152
Assembly language, 243
Assembly language programming, 152
Astable multivibrator, 79
Automatic soldering, 8

Base, 27
BASIC, 243
BCD counter, 98, 101
BCD-7-segment decoder, 96, 102
Beam deflection system, 108
Beam focusing system, 106
Binary
  adder, 73
  addition, 41
  coded decimal, 34
  comparator, 57
  counter, 98
  division, 43
  down counter, 100
  multiplication, 43
  number system, 28
  subtraction, 42
  up counter, 100
  words, 30

Bi-polar process, 11
Bistable multivibrator, 65, 79
Biquinary code, 37
Bits, 28
Blanking bits, 116
Board construction, 3
Boolean equations, 58, 137
Bubble memory. *See* Memories
Bus, 150
Byte, 30

'Can' pack, 17
Cards, 20
Central processor unit (CPU), 150, 153
Centralized instrument warning system, 146
Circuit diagram symbols, 240
Clock signal, 66
Clock oscillator, 69
COBOL, 243
Convergence card, 119
Coding, 33
Collector-diffused isolation, 11
Colour
  CRT, 109
  decoder, 113
  generation, 112
  mixing, 113
Column decoder. *See* Memories
Combined gate functions, 138
Combinational logic circuits, 49, 70, 73
Compacted mode, 129
Comparator, 79
Compiler, 153
Complement state, 65
Computers
  analog, 149
  applications, 164
  architecture, 150
  capacity, 152
  classification, 152
  digital, 150
  hardware, 150
  highway, 150
  languages, 152, 243
  organization, 150
  software, 150, 171
Cone of confusion, 203
Contrast ratio, 89

Control bus, 151
Counters, 69, 112, 203
Counting, 97
'Cross talk', 187
Cathode ray tube (CRT)
  colour, 109
  displays, 105, 167, 174
  principle, 106

Data
  alphanumeric, 113
  base, 180
  base loader, 180
  bus, 150, 154, 155
  highway, 154, 206
  pages, 181, 184
  storage, 156
  transfer, 153
  transmission, 44
Decimal number system, 27
Decoders, 70, 85, 155, 163
Decoding, 93
Demultiplexer, 72, 113, 207
Differential amplifier, 74
Digital air data computer, 164
Digital computer, 150
Digital-to-analog converter, 153, 163, 207, 210
Diode matrix, 159
Direct memory access, 152
Displays, 81
Divide-by-3600 counter, 203
Dot-matrix, 87, 209
D-type flip-flop. *See* Flip-flops
Dual functions of gates, 60
Dual-in-line pack, 17, 61
Dynamic RAM. *See* Memories
Dynamic scattering, 90

ECAM, 129
EICAS, 126
8-4-2-1 code, 34, 98
EFIS, 117
Electrically-alterable ROM. *See* Memories
Electromagnetic recording, 195
Electron-beam lithographic, 13–15
Electron-beam scanning, 105
Electron gun, 106, 112, 120

Index 273

Electronic displays, 105
Electronic instrument display systems, 117
Electrostatic discharge sensitivity, 213
Electrostatic-sensitive devices, 213
Encoders, 70, 163
Encoding, 93
Encoding altimeter, 192
Encoding disc, 193
Engine 'signature', 199
Entertainment services, 206
Epitaxy, 16
Epitaxial layer, 16
Erasable PROM. *See* Memories
Etching process, 7
Excess 3 code, 35
Exclusive NOR gate, 57
Exclusive OR gate, 57
Exponent, 27

Failure annunciation, 125
Failure-related mode, 131
Feedback, 76
Field-effect LCD, 90
Film circuits. *See* Integrated circuits
5-4-2-1 Code, 36
Five-bit codes, 36
Flat pack, 17
Flexible printed circuits, 8
Flight data acquisition system, 199
Flight data recording, 193
Flight management systems, 123, 171, 173, 179, 244
Flight phase-related mode, 131
Flip–flops
  applications, 67, 141, 146, 156
  complement logic state, 65
  D-type, 66
  fundamental operation, 65
  J–K type, 66, 67, 98
  latch, 66
  master–slave, 66
  normal logic state, 65
  S–R type, 65
FORTRAN, 243
Frequency divider, 103
Frequency division multiplexing, 71
Free-running multivibrator, 69
Full-adder, 73

Gadolinium–gallium garnet, 161
Gas discharge display, 91
Gray code, 38, 193
Grounding wrist strap, 218

Half-adder, 73
Handling of devices, 213
Hardware, 150
'Hard wiring', 3
Hexadecimal number system, 32
Higher-level language, 152
HSI display modes, 123
Hybrid integrated circuit. *See* Integrated circuits
Hybrid computer, 152

Hybrid logic, 48

Identification decals, 216
ILS mode, 125
Index, 27
Inhibited gate, 54
Input/output ports, 151
Instruction cycle, 151
Integrated circuits
  'can' pack, 17
  dual-in-line, 17
  epitaxial layer, 16
  fabrication, 12, 18
  flat pack, 17
  gold-wire bonding, 20
  hybrid, 12, 18
  laser beam trimming, 19
  monolithic, 11
  multichip, 12
  photomasks, 13
  printing pastes, 19
  reticle, 13
  scale of integration, 12
  substrate, 10, 15
  thick-film, 10
  thin-film, 10
  pack coding, 228
Integrating amplifier, 77
Interrogation modes, 189
Interrupts, 152
Inverter, 53
Inverting input, 74
Inverting logic, 52

J–K flip-flop. *See* Flip–flops
Johnson code, 37

Karnaugh maps, 59
Kilobits, 157

Label, 155
Landing gear aural warning system, 144
Large-scale integration, 12
Latch. *See* Flip–flops
Least significant bit, 102
Least significant digit, 27
Light emitting diodes, 85
Liquid crystal display, 88
Logic
  circuit equations, 58, 137
  devices, 65
  diagrams, 135, 141
  gates, 47
  gate data, 62
  hybrid, 48
  negative, 48, 60
  positive, 48, 60
  symbols, 50
Low pressure warning system, 141

Machine
  code, 152
  language, 152
  language program, 152

Maintenance mode, 129
MAP mode, 124
Master diagrams, 5, 12
Master–slave flip-flop. *See* Flip–flops
Matrix, 111
Matrix display, 82
Medium-scale integration, 12
Memories
  access time, 156
  addressable locations, 157
  bubble, 161, 180
  capacity, 157
  column decoder, 158
  diode matrix, 159
  dynamic RAM, 157
  electrically-alterable ROM, 160
  erasable PROM, 160, 180
  matrix, 115, 157
  non-volatile, 157, 162, 176
  programmable ROM, 160
  random-access (RAM), 157
  read-only (ROM), 159
  read/write, 157
  row decoder, 158
  static RAM, 157
  volatile, 157
Microprocessors, 153, 203, 209
Mnemonic instructions, 152
Mode annunciator, 178
Modularized circuit, 21
Modules, 20
Monolithic integrated circuit. *See* Integrated circuits
Monostable multivibrator, 69, 79
MOS, 11
MOS register, 69
Most significant bit, 102
Most significant digit, 27
Multichip integrated circuit. *See* Integrated circuits
Multiplexer, 70, 207, 209
Multiplexing, 70, 197, 206

NAND gate, 54
Nanoseconds, 156
Navigation data base, 173
Negated, 54
Negative logic, 48, 60
Non-inverting input, 74
Non-volatile memory. *See* Memories
NOR gate, 56
Normal state, 65
NOT logic, 52
Number identification, 28
Number systems, 27

Octal number system, 30
Offset, 80
One-of-eight decoder, 159
Ones complement, 43
Operand, 150
Operation code, 150
Operational amplifiers, 73, 208
Optoelectronic display, 81

274   Index

OR gate, 52
OR matrix, 137

PAFAM system, 167
Parity bits, 40
Parity check, 40
Parity word, 41
PASCAL, 243
Passive display, 81
Performance advisory system, 171
Performance data computer system, 173
Phase-locked loop, 203, 209
Photolithographic process, 7, 15
Photomasks. *See* Integrated circuits
'Pink-poly', 217
PLAN mode, 125
Plasma panel, 91
Positive logic, 48, 60
Power, 27
Practical logic diagrams, 141
Precision rectification, 76
Pressure error, 166
Primary engine parameters, 127
Printed circuits, 2
Printing pastes. *See* Integrated circuits
Programmable ROM. *See* Memories
Protection against static, 215
Pulse-code modulation, 207

Qualitative display, 81
Quantitative display, 81

Radials, 200
Radix, 27
Random access. *See* Memories
RAM. *See* Memories
Raster scanning, 109, 115
ROM. *See* Memories
Read/write. *See* Memories
Reflective code, 38

Remote light sensor, 120
Repetition rate, 155
Resist, 7
Reticle. *See* Integrated circuits
Return echoes, 110
Rho–theta form, 109
Ring-counter, 69
Ring-counter code, 37
Ripple counter, 103
'Rolling digit', 211
Row decoder. *See* Memories
S–R flip–flop. *See* Flip–flops

Sampling rates, 199
Scale of integration, 12
Scan conversion, 110
Scan generator, 96
Scratch pad, 185
Screen format, 110
Secondary engine parameters, 127
7-4-2-1 Code, 36
7-Segment matrix, 82
Serial access, 157
Servo-operated instruments, 207
Sequencer, 69
Sequential logic circuit, 65, 67, 69
Shadow mask, 112
Shift-counter code, 37
Shift register, 67, 156, 203
'Shunts', 217
Signal function equations, 139
Sign/status matrix, 156
Slewing rate, 80
Small-scale integration, 12
Software, 150, 171
Source/destination identifier, 156
Special-purpose computer, 152
'Starburst', 86
State indicator, 54
Static bleed-off time, 215
Static charges, 214

Static electricity, 213
Static-free packaging, 216
Static-free work stations, 217
Static RAM. *See* Memories
Status messages, 133
Storage register, 168
Stroke pulse method, 115
Substrate, 10, 15
Summing amplifier, 77
Symbol generator, 118, 120, 121, 169
Synchronous counter, 103

Thick-film circuit. *See* Integrated circuits
Thin-film circuit. *See* Integrated circuits
Time delay, 147
Time division multiplexing, 71, 206
Transponder, 189
Truth tables 51, 97
Twisted nematic, 90
Twos complement, 43

UV-PROM, 180

Very-large-scale integration, 12
Video/monitor card, 119
Volatile memory. *See* Memories
Voltage-following amplifier, 77
VOR mode, 125
VOR system, 200

Weather radar, 105, 109
Weight, 27
Wired gates, 57
Writing head, 195

$X$-axis fields, 109
$X$–$Y$ coordinate format, 109

$Y$-axis fields, 109
Yttrium–iron garnet, 161